Transistor
Circuit
Approximations

OTHER BOOKS BY ALBERT PAUL MALVINO

Electronic Principles
Digital Computer Electronics
Digital Principles and Applications (with D. Leach)
Resistive and Reactive Circuits
Electronic Instrumentation Fundamentals
Experiments for Transistor Circuit Approximations
Experiments for Electronic Principles (with G. Johnson)

Transistor Circuit Approximations
Third Edition

Albert Paul Malvino, Ph.D.

Foothill College
Los Altos Hills, California

West Valley College
Saratoga, California

Gregg Division
McGraw-Hill Book Company

New York	Atlanta	Dallas
St. Louis	San Francisco	Auckland
Bogotá	Düsseldorf	Johannesburg
London	Madrid	Mexico
Montreal	New Delhi	Panama
Paris	São Paulo	Singapore
Sydney	Tokyo	Toronto

Sponsoring Editor: George J. Horesta
Editing Supervisor: Linda Stern
Design Supervisor: Tracy A. Glasner
Production Supervisor: Frank P. Bellantoni

Text Designer: Denis Sharkey
Cover Designer: Al Manso
Technical Studio: Burmar Technical Corporation

Library of Congress Cataloging in Publication Data

Malvino, Albert Paul.
  Transistor circuit approximations.

  Includes index.
  1. Transistor circuits.   I. Title.
TK7871.9.M317   1980   621.3815'3'0422   79–18580
ISBN 0-07-039878-X

Transistor Circuit Approximations, Third Edition

  2 3 4 5 6 7 8 9 0 DODO 8 8 7 6 5 4 3 2 1 0

TO MY WILD IRISH ROSE

It is the mark of an instructed mind
to rest satisfied
with that degree of precision
which the nature of the subject admits,
and not to seek exactness
where only an approximation
of the truth is possible

ARISTOTLE

Contents

Contents

point   9-4. Power formulas   9-5. Large-signal gain and impedance   9-6. Transistor power ratings

Contents

Preface

The philosopher René Descartes divided every problem into as many parts as possible. Then, he analyzed the simplest part before proceeding to the next more difficult item. Going from the simple to the complex is a key idea in the scientific method of Descartes.

Transistor Circuit Approximations, Third Edition, is an example of the Descartes approach. It starts with ideal approximations (the simplest) and proceeds to second and third approximations (the more complex). With the ideal-transistor approximation described in this third edition, you can sail through problems and arrive at an ideal solution. Then, when necessary, higher approximations can improve accuracy. This idealize-and-improve approach is the most efficient way to solve the problems encountered in industry.

In revising *Transistor Circuit Approximations* for this third edition, I read hundreds of questionnaires completed by instructors throughout the United States. More than 75 percent of these questionnaires asked for the following to be included: power supplies, feedback, linear ICs, and pulse and digital circuits. More coverage of multistage amplifiers, push-pull power amplifiers, types of coupling, JFETs, and MOSFETs was also requested. To satisfy these requirements, I have rewritten virtually every word in the book, cutting the obsolete material and emphasizing the bread-and-butter topics.

This new edition is shorter than the previous one and should fit better into a quarter or semester introduction to semiconductor devices. The bipolar diode and transistor still dominate the early chapters, but now the later chapters expand the discussion of JFETs and MOSFETs, especially in switching applications. Also new in the book is a chapter on operational amplifiers.

Transistor Circuit Approximations is intended for a course in electronic

Preface

devices and circuits usually offered in technician training programs found in two-year colleges, technical institutes, and industrial training programs. A prerequisite course in direct and alternating current circuits is assumed. Previous editions have used conventional flow in all circuit analysis. Because of many requests, this third edition has been written in such a way that instructors can employ either conventional flow or electron flow in classroom discussions.

As before, each chapter contains a glossary of the most important terms to help the reader learn the language of basic electronics. To improve reader participation, each chapter contains a self-testing review that reinforces the important ideas. In addition, a set of problems appears at the end of each chapter.

A correlated laboratory manual, *Experiments for Transistor Circuit Approximations,* is also available. In conjunction with the textbook, this manual gives a student "hands-on" experience that clears up difficult topics and rounds out understanding.

I owe special thanks to Bobby L. Alexander of Texas State Technical Institute, L. E. Courtney of Midlands Technical College, Jack T. Simpson of ITT Technical Institute, and Norman H. Sprankle of Humboldt State University. They reviewed the manuscript for this third edition. I found their suggestions to be invaluable in showing me how to enhance the appeal, content, and readability of this book.

Albert Paul Malvino

Chapter 1

Semiconductor Physics

This chapter is about semiconductors, free electrons, holes, and doping. Learn this material, because it will help you understand how solid-state devices work.

1-1 GERMANIUM AND SILICON ATOMS

An isolated atom of germanium contains a nucleus with 32 protons. When this atom is electrically neutral, 32 electrons orbit the nucleus as shown in Fig. 1-1a. The first orbit (also called a *shell*) contains 2 electrons; the second orbit has 8 electrons; the third has 18; and the fourth contains 4. It is the 4 electrons in this outer orbit, often called the *valence* orbit, that makes germanium an important semiconductor material.

A silicon atom has a nucleus with 14 protons. When the atom is electrically neutral, 14 electrons are in orbit around the nucleus. As shown in Fig. 1-1b, the first shell has 2 electrons; the second shell has 8 electrons; and the valence shell has 4 electrons. Again, it is the 4 electrons in the valence shell that make silicon a practical semiconductor material.

To emphasize the importance of the valence electrons, we will simplify

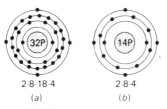

2-8-18-4   2-8-4
   (a)   (b)

Figure 1-1 *(a)* Germanium. *(b)* Silicon.

1

the atomic diagrams as indicated in Fig. 1-2. The inner part of each atom is known as a *core;* it contains the nucleus and all inner-orbit electrons. As you see, each outer shell has 4 electrons. This is why germanium and silicon are classified as *tetravalent* elements (*tetra* is Greek for "four").

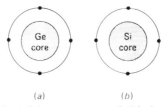

(a)　　　　(b)

Figure 1-2 Valence shell of *(a)* germanium and *(b)* silicon.

For reasons to be given later, germanium devices are virtually obsolete. Silicon, on the other hand, has become the most prominent semiconductor material in use today. Almost all diodes, transistors, and integrated circuits are made out of silicon. Unless otherwise indicated, all devices in this book are silicon.

1-2 CRYSTALS

When isolated silicon atoms combine to form a solid, they arrange themselves in an orderly three-dimensional pattern called a *crystal.* The forces holding the atoms together are known as *covalent bonds.* What follows is a brief description of covalent bonding.

An isolated silicon atom has four valence electrons. For reasons covered in advanced chemistry and physics, silicon atoms combine so as to have *eight electrons in the valence orbit.* To manage this, each silicon atom positions itself between four other silicon atoms (see Fig. 1-3a). Each neighbor shares an electron with the central atom. In this way, the central atom has picked up four electrons, making a total of eight in its valence orbit. Actually,

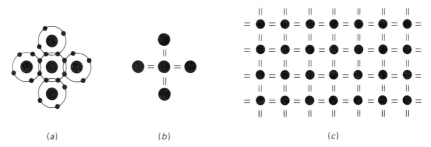

(a)　　　　　(b)　　　　　　　(c)

Figure 1-3 *(a)* Covalent bonds. *(b)* Bonding diagram. *(c)* Crystal diagram.

the electrons no longer belong to a single atom; they are shared by adjacent atoms. It is this sharing that sets up the covalent bond.

Figure 1-3*b* symbolizes the mutual sharing of electrons. Each line represents a shared electron. Each shared electron establishes a bond between the central atom and a neighbor. For this reason, we call each line a covalent bond.

Here's one way to understand covalent bonding. In Fig. 1-3*b,* the positive cores attract each pair of shared electrons with equal and opposite forces. The motionless pulling on the shared electrons is what holds the atoms together. It's analogous to tug-of-war teams pulling on a rope. As long as the teams pull with equal and opposite force, both teams remain immobile and bonded together.

Figure 1-3*c* shows the bonding diagram for many silicon atoms in a crystal. Because of the shared electrons, each atom has eight electrons in its valence shell. And because of the core attraction, these eight electrons are tightly held within the valence orbit.

1-3 CONDUCTION IN PURE SILICON

How well does pure silicon conduct? When the temperature is at absolute zero ($-273°$C), pure silicon has no free electrons, as shown in Fig. 1-4*a,* because all valence electrons are tightly held by the positive cores. Therefore, at absolute zero temperature a silicon crystal acts like a perfect insulator.

As the *ambient* (surrounding) temperature increases, however, *thermal* (heat) energy enters the crystal and causes the silicon atoms to vibrate about their equilibrium positions. The greater the temperature increase, the greater the vibration. This mechanical motion dislodges a few valence electrons, sending them into much *larger* orbits called *conduction-band* orbits. Once an electron is in a conduction-band orbit, the attraction of the cores is so weak that the electron is more or less free to move throughout the crystal. This is why conduction-band electrons are known as *free* electrons.

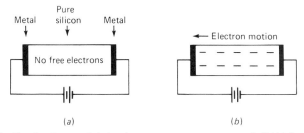

Figure 1-4 Conduction at *(a)* absolute zero temperature and *(b)* higher temperatures.

Figure 1-4*b* indicates the free electrons by minus signs. Since the battery applies a difference of potential to the ends of the crystal, the free electrons move toward the left, setting up a current. This current is small, because only a few free electrons are produced by thermal energy.

If the temperature is increased further, more free electrons appear and the current increases. At room temperature (around 25°C) the current is too small for silicon to be classified as a conductor and too large for silicon to be called an insulator. This is the reason for referring to silicon as a *semiconductor.*

1-4 HOLES

Figure 1-5 shows a silicon atom with four neighboring atoms. When an electron is knocked out of a valence orbit, it becomes a free electron and can move to another part of the crystal. The vacancy left in the outer shell is called a *hole.* This hole acts like a positive charge because it will attract and capture any free electron that happens to drift into the immediate vicinity.

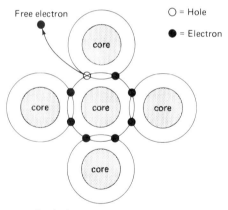

Figure 1-5 Production of a hole.

Recombination and lifetime

Thermal energy produces an equal number of free electrons and holes in a pure silicon crystal. The free electrons move randomly throughout the crystal. Occasionally, a free electron passes close enough to a hole to be attracted and captured. When this happens, the hole disappears and the free electron becomes a bound electron. This merging of a free electron and a hole is known as *recombination.*

At any instant the following takes place inside the crystal:

1. Thermal energy is continuously producing free electrons and holes.
2. Other free electrons and holes are recombining.
3. Some free electrons and holes exist in an in-between stage; they were previously generated but have not yet recombined. The average time these free electrons and holes exist before recombining is called the *lifetime*.

Hole current

Two components of current are possible in a semiconductor: the movement of free electrons is one component, and the movement of holes is the other. The flow of free electrons is already familiar; it's the kind of current that exists in a copper wire. The idea is this. The free electrons travel in large conduction-band orbits and move easily from one atom to the next.

Hole current is different. Figure 1-6 shows why. Notice the hole at the extreme right. Adjacent to this hole is a bound electron at position *A*. Being in the same orbit, this bound electron can easily move into the hole.

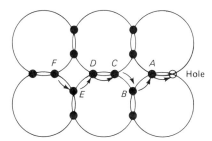

Figure 1-6 Hole movement.

When it does, a new hole appears at position *A*. The new hole at *A* can now attract and capture the bound electron at position *B*. When this happens, a new hole appears at *B*. This action can continue with a bound electron moving along the path shown by the arrows. In this way, a hole can move from one atom to another. Especially important, the hole moves opposite to the direction of the bound electrons because holes act like positive charges.

Two kinds of flow

The battery of Fig. 1-7 applies a difference of potential to the silicon crystal. The plus signs represent the holes and the minus signs represent the free electrons. Because of the battery polarity, the free electrons move to the left and the holes move to the right. Some recombination of free electrons and holes occurs, but thermal energy continuously produces new free electrons

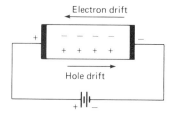

Figure 1-7 Components of current.

and holes. Therefore, some free electrons and holes always exist and contribute to the current.

The drift of free electrons to the left is one component of the current. When the free electrons reach the left end of the crystal, they enter the metal and pass on to the positive terminal of the battery. In the meantime, the negative battery terminal injects free electrons into the right side of the crystal, maintaining a continuous flow of free electrons.

The flow of holes to the right is the second component of current in Fig. 1-7. When these holes reach the right end of the crystal, free electrons recombine with these holes and become bound electrons. In the meantime, thermal energy keeps creating new holes at the left end of the crystal. In this way, a continuous flow of holes exists.

A final point should be emphasized. Inside the crystal, the current carriers are free electrons and holes. The free electrons travel in *large* conduction-band orbits; the holes travel in *small* valence-band orbits. It's analogous to the flow of traffic on a two-tier bridge with free electrons on the upper level and holes on the lower level.

1-5 EXTRINSIC SEMICONDUCTORS

The number of free electrons and holes produced by thermal energy is too small to be of practical use. But it is possible to increase the current carriers by *doping*. A semiconductor that has been doped is called an *extrinsic* semiconductor; if undoped, it is called *intrinsic*.

Increasing the free electrons

A *pentavalent* element like arsenic has five valence electrons in each atom. One way to dope silicon is to melt it (this breaks the covalent bonds) and to add some arsenic atoms. These pentavalent atoms diffuse through the molten silicon. Later, when the silicon cools, a solid crystal forms. Once again, each atom in the crystal has four neighboring atoms that share electrons. Most of the atoms are silicon, but occasionally we find an arsenic atom.

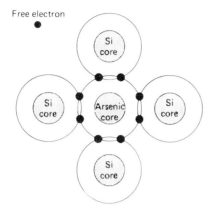

Figure 1-8 An *n*-type semiconductor.

In Fig. 1-8, the arsenic atom has eight electrons in its valence shell. This atom originally had five valence electrons. Since each neighbor donates one electron, an extra electron is left over. This extra electron becomes a free electron. In other words, each arsenic atom produces one free electron. By varying the amount of arsenic added to the molten silicon, we can control the number of free electrons.

Majority and minority carriers

The creation of free electrons by doping is different from the thermal production of free electrons and holes. Only free electrons are produced by doping with arsenic. As a result, there are more free electrons in the doped silicon than holes. The few holes present are the result of thermal energy.

A pure semiconductor that has been doped with a pentavalent element is called an *n*-type semiconductor. The *n* stands for "negative," referring to the excess of free electrons. Since an *n*-type semiconductor has more free electrons than holes, we call the free electrons the *majority* carriers and the holes the *minority* carriers.

Increasing the holes

Elements like aluminum, boron, and gallium are *trivalent* (three valence electrons). When used to dope silicon, trivalent elements produce holes. For instance, Fig. 1-9 shows an aluminum atom with four neighboring silicon atoms. Originally, the aluminum atom had three valence electrons. Since each neighbor shares one electron, the valence shell contains only seven electrons. This is why a hole appears. Since each aluminum atom produces

7

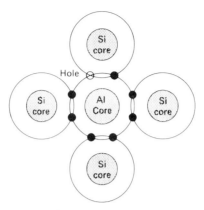

Figure 1-9 A p-type semiconductor.

one hole, we can control the number of holes by varying the amount of aluminum.

A semiconductor doped with trivalent impurities is called a p-type semiconductor. The p stands for "positive," referring to the excess of holes. Since a p-type semiconductor has more holes than free electrons, the holes are the majority carriers and the free electrons are the minority carriers. Figure 1-10 summarizes the two types of doped semiconductors.

Figure 1-10 Majority and minority carriers.

1-6 SEMICONDUCTOR DEVICES

By combining p and n materials in different ways, we get the semiconductor devices used in modern electronics. The remaining chapters discuss these devices and their applications in detail. For now, here is a brief description of the basic semiconductor devices.

The diode

Figure 1-11a shows the *diode;* it combines p and n materials. The diode is the most basic *solid-state* (semiconductor) device. It lets majority carriers flow easily only in one direction. In other words, it is like a one-way street.

8

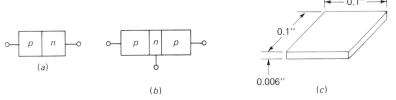

Figure 1-11 *(a)* Diode. *(b)* Transistor. *(c)* Chip.

Because of this one-way action, the diode is used in *power supplies,* circuits that convert ac line voltage to dc voltage suitable for driving electronics equipment.

The transistor

Figure 1-11*b* is a *transistor,* a device with three doped regions. Shockley worked out the theory of the transistor in 1949, and the first transistor was produced in 1951. This basic device has the ability to amplify weak signals. Because of this, the transistor's impact on electronics has been enormous. Besides starting the multibillion-dollar semiconductor industry, the transistor has spawned all kinds of related inventions like integrated circuits, optoelectronic devices, and microprocessors.

The integrated circuit

Figure 1-11*c* shows a *chip,* a small piece of semiconductor material. The dimensions are representative; chips are often smaller than this, occasionally larger. By advanced photographic techniques, a manufacturer can produce circuits on the surface of this chip, circuits with many diodes, resistors, transistors, etc. The finished network is so small you need a microscope to see the connections. We call a circuit like this an *integrated circuit* (IC).

A *discrete circuit* is different. It is the kind of circuit you build when you connect separate resistors, capacitors, transistors, etc. Each component or device is discrete, that is, distinct or separate from the others. This differs from an integrated circuit, where the components are atomically part of the semiconductor chip.

GLOSSARY

crystal A solid whose atoms form a repeating geometric pattern. In a silicon crystal, each atom forms covalent bonds with four neighboring atoms.

9

Transistor Circuit Approximations

doping Adding pentavalent or trivalent elements to a pure semiconductor to increase the number of free electrons or holes.

extrinsic semiconductor Doped germanium or silicon.

hole A vacancy in the outer shell, produced by thermal energy or by doping.

intrinsic semiconductor Pure germanium or silicon. The only current carriers are the free electrons and holes produced by thermal energy.

lifetime The average amount of time a free electron or hole exists before recombining.

***n*-type semiconductor** A semiconductor that has been doped to get an excess of free electrons.

***p*-type semiconductor** A semiconductor that has been doped to get an excess of holes.

recombination The merging of a free electron and hole.

semiconductor A material like germanium or silicon whose electrical properties are between those of an insulator and a conductor.

SELF-TESTING REVIEW

Read each of the following and provide the missing words. Answers appear at the beginning of the next question.

1. The two most important semiconductor materials are __Si__ and __Ge__. Isolated atoms of these materials have __4__ electrons in the valence shell.

2. *(germanium, silicon, four)* A __crystal__ results when silicon atoms combine into a solid piece of material. In this structure, each atom has four neighbors which share their electrons to produce a total of __8__ electrons in the valence shell of each atom.

3. *(crystal, eight)* At absolute zero temperature, pure silicon acts like an __insulator__ because no free electrons exist in the semiconductor. Above absolute zero, __thermal__ energy dislodges some electrons from valence shells.

4. *(insulator, thermal)* A __hole__ is a vacancy in the outer shell. It can attract and capture a nearby electron. This merging is called __recombination__

5. *(hole, recombination)* Adding pentavalent or trivalent impurities to a pure

10

semiconductor is called _doping_ . The *n*-type semiconductor has an _excess_ of free electrons, and the *p*-type has an excess of _holes_ .

6. *(doping, excess, holes)* The majority carriers in an *n*-type semiconductor are the _free_ electrons, and the minority carriers are the _hole_ . In a *p*-type semiconductor, the _majority_ carriers are holes and the minority carriers are free electrons.

7. *(free, holes, majority)* _lifetime_ is the average amount of time that holes and free electrons exist before recombining.

8. *(Lifetime)* Pure silicon is called an intrinsic semiconductor. Doped silicon is an extrinsic semiconductor.

Chapter 2

Rectifier Diodes

This chapter is about the *rectifier diode,* a device that lets carriers flow only in one direction. Included in the discussion are approximations that simplify the analysis of diode circuits.

2-1 THE *pn* JUNCTION

At room temperature, *p* material has holes produced by doping and only a few free electrons produced by thermal energy. On the other hand, *n* material has mostly free electrons and only a few thermally produced holes.

The carriers

Figure 2-1*a* shows *p* material. The plus signs are the holes and the circled minus signs are the atoms the holes are in. If a hole moves away from its atom, the atom becomes negatively charged. This negative charge is immobile; it cannot move because the atom is part of the crystal structure. There-

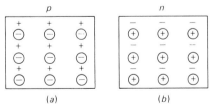

Figure 2-1 Current carriers and immobile atoms. *(a) p*-type. *(b) n*-type.

fore, in Fig. 2-1*a* the positive charges are free to move, but the negative charges are not.

Similarly, Fig. 2-1*b* shows *n* material. Here, the minus signs symbolize free electrons, and the plus signs are the atoms the free electrons orbit. If a free electron moves away from its atom, it leaves a positive charge behind. This positive charge is not free to move. In *n* material, therefore, the negative charges are free to move, but not the positive charges.

Depletion layer

A manufacturer can produce a crystal with *p* material on one side and *n* material on the other, as shown in Fig. 2-2*a*. At the instant the junction between the two materials is formed, all the holes are still in the *p* material and all the free electrons are in the *n* material. The holes and free electrons tend to move randomly in all directions. Some of the holes and free electrons will move across the junction and recombine.

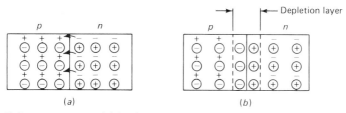

(a)   (b)

Figure 2-2 *p-n* junction. *(a)* At the instant of formation. *(b)* Depletion layer.

This recombination produces a narrow region at the junction called the *depletion layer* (see Fig. 2-2*b*). Because of the recombination, the depletion layer has no holes or free electrons; it contains only positively and negatively charged atoms as shown.

Barrier potential

As the depletion layer builds up, a difference of potential appears across the junction because of the negative charges on the left and the positive charges on the right. Eventually, this difference of potential becomes large enough to prevent hole and free-electron movement across the junction.

The difference of potential at the junction is called the *barrier potential.* At room temperature (around 25°C) the barrier potential is approximately 0.3 V for germanium and 0.7 V for silicon.

The barrier potential changes with temperature. For germanium and silicon

13

devices, the barrier potential decreases approximately 2.5 mV for each Celsius degree rise. As a formula, the change in barrier potential is

$$\Delta V = -0.0025 \ \Delta T \qquad \textbf{(2-1)}$$

where Δ stands for "the change in."

2-2 FORWARD AND REVERSE BIAS

A *pn* crystal acts like a *diode,* a device that allows carriers to flow easily only in one direction. (It's like a one-way street.) To understand the action of a diode, look at Fig. 2-3. The barrier potential tries to stop holes and free electrons from moving across the junction. The battery, however, is repelling holes and free electrons toward the junction. When the battery potential is large enough to overcome the barrier potential, a continuous current can exist because holes and free electrons will cross the junction in large numbers.

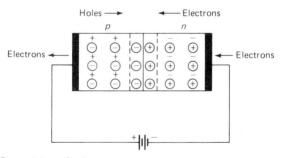

Figure 2-3 Current in a diode.

The path of one electron

For a better understanding of the action, let us follow one electron around the entire circuit of Fig. 2-3. To begin with, a free electron leaves the negative battery terminal and enters the *n* material. Once inside the *n* material, this free electron moves toward the junction. Somewhere in the vicinity of the junction, this free electron recombines with a hole and becomes a valence electron. As a valence electron, it continues moving to the left, traveling through holes in the *p* material. When this electron reaches the left end of the crystal, it enters the wire and flows on to the positive battery terminal.

An equivalent viewpoint

As discussed in the preceding chapter, the motion of a valence electron to the left is equivalent to a hole moving to the right. In Fig. 2-3, therefore, here is how we can visualize the current. The battery forces holes in the *p* material and free electrons in the *n* material to move toward the junction, where they recombine. The battery is continuously producing new holes at the left end of the diode and injecting free electrons into the right end. Because of this, we get a continuous flow of majority carriers toward the junction. In other words, we get a large current through the diode called the *forward current*.

Reversing the battery

What happens if we reverse the battery as shown in Fig. 2-4? In this case, the battery potential aids the barrier potential in preventing majority carriers from crossing the junction. For this reason, the current drops to a very small value known as the *reverse current*.

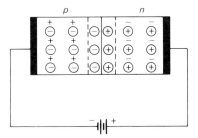

Figure 2-4 Reverse bias.

What causes reverse current? Because of thermal energy, the *n* material has a few holes and the *p* material has a few free electrons. The battery repels these minority carriers toward the junction, where they recombine. Since a few minority carriers are being continuously produced by thermal energy, we get a continuous flow of these minority carriers through the diode.

Besides the minority carriers, a few carriers can flow along the surface of the crystal. This small unwanted current is known as *surface leakage*. It is directly proportional to the applied voltage.

But here is the point to remember. The reverse current produced by minority carriers and surface leakage is so small that we can neglect it in most applications.

Summary

Here are the main points about diode action:

1. Where the applied voltage overcomes the barrier potential (the *p* side is more positive than the *n* side), the current produced is large because majority carriers cross the junction in large numbers. This condition is called *forward bias.*
2. When the applied voltage aids the barrier potential (*n* side more positive than *p* side), the current is small. This state is known as *reverse bias.*

2-3 THE DIODE CURVE

Figure 2-5 shows the schematic symbol of a diode. Current is large in the forward direction, but small in the reverse direction. As a reminder of this one-way action, the diode arrow points in the easy direction of *conventional current.* If you prefer using *electron flow,* the easy direction is against the diode arrow.

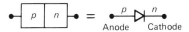

Figure 2-5 Schematic symbol of a diode.

Knee voltage

Figure 2-6*a* shows an adjustable source across the diode. When the applied voltage is zero, no current exists. As we increase the voltage, carriers begin to flow. The current increases slowly at first. When the applied voltage is large enough to overcome the barrier potential, the current increases rapidly.

Figure 2-6*b* summarizes the relationship between diode current and voltage. Notice how the current is small until we reach V_K. Voltage V_K is called the *knee voltage;* it is the voltage at which the current sharply increases. Knee

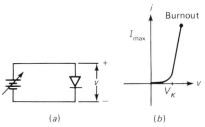

(a)   (b)

Figure 2-6 *(a)* Forward bias. *(b)* Forward characteristic.

16

voltage is approximately equal to the barrier potential. In other words, V_K is around 0.3 V for germanium diodes and 0.7 V for silicon diodes.

Maximum forward current

There is a limit to the amount of current a diode can handle. When we increase the applied voltage well beyond the knee voltage, we eventually reach a current large enough to destroy the diode. For instance, the 1N4001 is a commercially produced diode with a maximum current of 1 A. Such a diode can tolerate up to 1 A of continuous forward current without being destroyed.

Breakdown

Figure 2-7a shows reverse bias. If the applied voltage is zero, no current exists. When we increase the voltage, very little current exists. But there is a limit to the reverse voltage a diode can withstand. If the voltage is too large, the diode will break down and the current will become large.

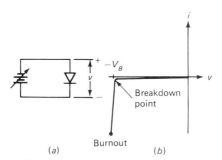

Figure 2-7 *(a)* Reverse bias. *(b)* Reverse characteristic.

Figure 2-7b summarizes the idea. Notice how the reverse current is extremely small until we reach V_B. Voltage V_B is called the *breakdown voltage;* it is the voltage where reverse current suddenly increases sharply. This breakdown of the diode is caused by either of two effects: *avalanche* or *zener.* In both cases, bound electrons are knocked out of valence shells and become free electrons.

Breakdown voltage varies from one diode type to another. For example, the 1N4001 has a breakdown voltage of 50 V, whereas a 1N4002 has a breakdown voltage of 100 V. Except for special diodes described in the next chapter, the applied reverse voltage across a diode should always be less than the breakdown value specified for the diode.

Transistor Circuit Approximations

Main points

Figure 2-8 shows the overall curve for a semiconductor diode. Remember these points:

1. When forward-biased, a diode needs only a few tenths of a volt to conduct heavily.
2. When the diode is reverse-biased, current is negligibly small until reaching the breakdown voltage, typically 50 V or more.

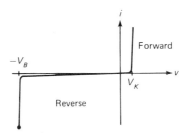

Figure 2-8 Diode curve.

2-4 THE IDEAL DIODE

To simplify the analysis of diode circuits, we will use approximations. The first (and simplest) approximation is the *ideal diode*. If a diode were ideal or perfect, it would have no forward voltage drop and no reverse current (see Fig. 2-9a). This is equivalent to a one-way switch that is closed when forward-biased and open when reverse-biased (Fig. 2-9b).

Extreme as the ideal-diode approximation seems at first, it gives adequate answers for most diode circuits. There will be times when the approximation is unacceptable, and for this reason, we will need a second and third approximation. But for all preliminary analyses of diode circuits, the ideal diode is superb.

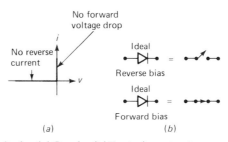

Figure 2-9 Ideal diode. *(a)* Graph. *(b)* Equivalent circuit.

18

EXAMPLE 2-1

Use the ideal-diode approximation to find the output waveform of Fig.
2-10a.

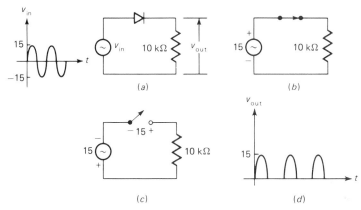

Figure 2-10

SOLUTION

The input is a sine wave with a positive peak of 15 V and a negative peak
of −15 V. During the positive half cycle, the diode is forward-biased because
conventional current is in the direction of the diode arrow (electron flow is
the opposite way). Figure 2-10b shows the circuit at the positive peak.
Because the switch is closed, the output has a positive peak of 15 V.

During the negative half cycle of input voltage, the diode is reverse biased.
Figure 2-10c shows the circuit at the negative peak. Because the switch is
open, no voltage reaches the output. To satisfy Kirchhoff's voltage law, 15
V appears across the diode.

We can summarize the action like this: each positive half cycle appears
across the output; each negative half cycle is blocked. This is why the output
is the *half-wave* signal shown in Fig. 2-10d. The diode circuit of Fig. 2-
10a is called a *half-wave rectifier.*

EXAMPLE 2-2

Calculate the peak current through the diode in Fig. 2-10a. $1.5mA$ Also, what is
the maximum reverse voltage the diode must be able to withstand? $15V$

SOLUTION

The peak current through the diode occurs at the positive peak (Fig. 2-10b)
and equals

19

$$I_{max} = \frac{15\ V}{10,000\ \Omega} = 1.5\ mA$$

The maximum reverse voltage that a diode must withstand during the cycle is called the *peak inverse voltage,* abbreviated PIV. This voltage occurs at the negative peak (Fig. 2-10c) and in this case equals 15 V.

2-5 THE SECOND APPROXIMATION

The ideal diode is the simplest but crudest approximation of a rectifier diode. The answers we get with this approximation give us an initial idea of how diode circuits work.

To improve the accuracy, we can include the forward voltage drop across the diode. A simple way to do this is to include the knee voltage. Figure 2-11a is the curve for the *second approximation;* this means voltage V_K is across the diode when it is conducting. The second approximation is equivalent to an ideal diode in series with a battery of V_K (Fig. 2-11b). In other words, we will allow 0.3 V for a germanium diode and 0.7 V for a silicon diode.

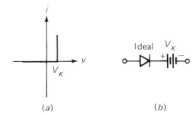

Figure 2-11 The second approximation: *(a)* Graph. *(b)* Equivalent circuit.

EXAMPLE 2-3

Use the second approximation to find the output waveform in Fig. 2-12a. Also, work out the maximum forward current and the peak inverse voltage.

SOLUTION

During the positive half cycle of input voltage, the first 0.7 V is wasted in overcoming the barrier potential; thereafter, the diode can conduct. Figure 2-12b shows the circuit at the positive peak. Kirchhoff's voltage law tells us the peak output voltage equals the source voltage minus the drop across the diode:

$$V_P = 15\ V - 0.7\ V = 14.3\ V$$

20

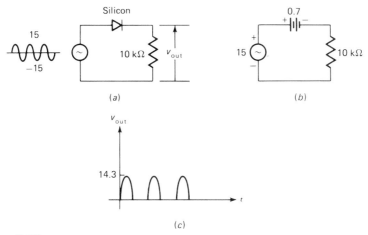

(a)

(b)

(c)

Figure 2-12

During the negative half cycle, the diode is open; therefore, no voltage reaches the output. For this reason, the final output waveform is a half-wave signal with a peak of 14.3 V as shown in Fig. 2-12c.

The maximum forward current occurs at the positive peak of input voltage and equals

$$I_{max} = \frac{14.3 \text{ V}}{10,000 \text{ } \Omega} = 1.43 \text{ mA}$$

What about the peak inverse voltage? When the diode is reverse-biased, it is open and no voltage appears across the output. Because of this, all of the source voltage is across the diode. Therefore,

$$PIV = 15 \text{ V}$$

2-6 THE THIRD APPROXIMATION

Still another approximation can be made for a rectifier diode to account more accurately for its forward voltage drop. The forward diode curve is not vertical above the knee; it has a slight upward slope to the right. This means more voltage is dropped across the diode as the current increases.

Graph

Figure 2-13a is the *third approximation* of a rectifier diode. This is more accurate than the previous approximations. The diode starts conducting

21

Transistor Circuit Approximations

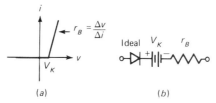

(a)   (b)

Figure 2-13 The third approximation. *(a)* Graph. *(b)* Equivalent circuit.

above V_K. Once conducting, it acts like a small resistor because the change in current is directly proportional to the change in voltage. This resistance is called the *forward* or *bulk* resistance of the diode and is symbolized by r_B.

Equivalent circuit

Figure 2-13b is the equivalent circuit for the third approximation, an ideal diode in series with a battery and a resistor. When the diode is conducting, the voltage across the third approximation is the sum of knee voltage plus the drop across the bulk resistance. In other words, it takes at least 0.3 V or 0.7 V to get conduction. The diode then drops additional voltage across r_B. The total voltage across the diode therefore is

$$V_F = V_K + I_F r_B \tag{2-2}$$

where $V_K = 0.3$ V for germanium diodes and 0.7 V for silicon diodes.

More on bulk resistance

A word or two on finding bulk resistance. If a curve tracer is available, you can display the forward diode curve. This curve will appear almost linear as shown in Fig. 2-14. By selecting any two points well above the knee, you can calculate the bulk resistance using

$$r_B = \frac{\Delta v}{\Delta i} \tag{2-3}$$

where Δv and Δi are the changes in voltage and current between the two points.

22

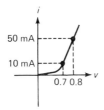

Figure 2-14 Estimating bulk resistance.

For instance, in Fig. 2-14 the change in voltage is 0.1 V and the change in current is 40 mA. Therefore, the bulk resistance is

$$r_B = \frac{0.1 \text{ V}}{40 \text{ mA}} = 2.5 \text{ }\Omega$$

EXAMPLE 2-4

Use the third approximation to calculate the maximum current and peak inverse voltage in Fig. 2-15a.

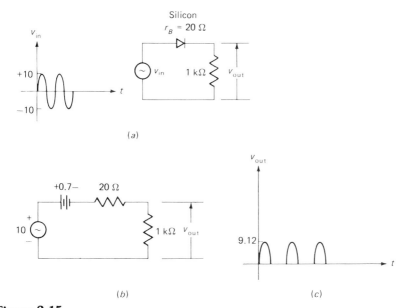

Figure 2-15

SOLUTION

Figure 2-15b shows the circuit at the positive peak of source voltage. The maximum forward current is

23

$$I_{max} = \frac{10\text{ V} - 0.7\text{ V}}{1000\ \Omega + 20\ \Omega} = \frac{9.3\text{ V}}{1020\ \Omega} = 9.12\text{ mA}$$

The peak voltage across the 1-kΩ resistor therefore is

$$V_p = 9.12\text{ mA} \times 1\text{ k}\Omega = 9.12\text{ V}$$

Figure 2-15c is the output waveform.

During the negative half cycle, the diode is open and all the source voltage appears across it. This means

$$\text{PIV} = 10\text{ V}$$

2-7 USING THE APPROXIMATIONS

The approximations can shorten the time required to analyze diode circuits; these approximations are also useful in transistor-circuit analysis. The ideal diode is the simplest approximation and the one we use most often. The second approximation is also used a lot. The third approximation is rarely used.

Here's a summary for the approximations:

1. In analyzing any rectifier-diode circuit, start with the ideal-diode approach. This gives a basic idea of how the circuit operates and is adequate for most cases where the source voltage is much greater than the knee voltage.
2. If source voltage is not large compared to knee voltage, use the second approximation. As a guide, if the peak source voltage is less than 10 V, you should include the effect of knee voltage by analyzing the circuit with the second approximation rather than the ideal diode.
3. If the circuit resistance in series with the diode is not large compared to bulk resistance, use the third approximation. As a guide, most silicon diodes have bulk resistances under 10 Ω, and typical silicon diodes have bulk resistances less than 1 Ω. Circuit resistance is usually much greater than bulk resistance, and this is why the third approximation is seldom used.

2-8 APPROXIMATING REVERSE CURRENT

A small current exists when a diode is reverse-biased. This implies the diode has a large resistance. This resistance, known as the *reverse resistance*, equals the reverse voltage divided by the reverse current. As a formula, this is expressed as

$$R_R = \frac{V_R}{I_R} \tag{2-4}$$

For instance, at 25°C a 1N4001 has a reverse current of 0.05 μA for a reverse voltage of 50 V. Therefore,

$$R_R = \frac{50 \text{ V}}{0.05 \text{ μA}} = 1000 \text{ M}\Omega$$

Reverse resistance changes with temperature. The higher the temperature, the greater the number of thermally-produced carriers and the lower the reverse resistance. As an example, at 100°C the 1N4001 has a reverse current of 1 μA for a reverse voltage of 50 V. This gives

$$R_R = \frac{50 \text{ V}}{1 \text{ μA}} = 50 \text{ M}\Omega$$

As you see, the reverse resistance of a 1N4001 decreases from 1000 MΩ to 50 MΩ when the temperature increases from 25°C to 100°C.

Silicon diodes are far superior to germanium diodes because they have much higher reverse resistance. A typical germanium diode has a reverse resistance of less than 1 MΩ, whereas the typical silicon diode has a reverse resistance much greater than 1 MΩ. Higher reverse resistance is one of the reasons that silicon diodes are preferred to germanium diodes.

EXAMPLE 2-5

The diode of Fig. 2-16a has a reverse resistance of 50 MΩ. What is the output waveform across the 1-MΩ resistor?

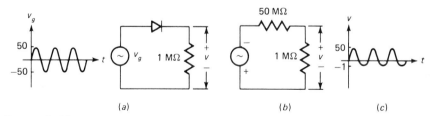

(a)   (b)   (c)

Figure 2-16 Effect of reverse resistance.

SOLUTION

If the diode were ideal, the output would be a half-wave signal. But since the load resistor is so large (1 MΩ), the reverse resistance will have some effect. Figure 2-16b shows the circuit during the negative half cycle of input

25

voltage. Since we have a voltage divider, the output is approximately 1/50 of the input voltage.

Figure 2-16c is the output waveform. Note that the positive peak is 50 V and the negative peak is −1 V. If this −1 V is unacceptable, either of two changes is possible. First, the 1 MΩ can be reduced to a lower value, like 100 kΩ. This results in a negative peak of 1/500 of the input voltage. Second, we can keep the load at 1 MΩ and use another diode type with a higher reverse resistance. For instance, a diode with a reverse resistance of 1000 MΩ means the negative peak is only 1/1000 of the input voltage.

GLOSSARY

avalanche At high enough reverse voltages, minority carriers can reach sufficient velocities to dislodge valence electrons. The newly freed electrons then reach high enough velocities to dislodge more electrons.

barrier potential The voltage across the *pn* junction produced by charged atoms in the depletion layer. It equals 0.3 V for germanium diodes and 0.7 V for silicon diodes.

breakdown voltage The reverse voltage where avalanche or zener effect begins to produce a large reverse current.

bulk resistance This is the ohmic resistance of the *p* and *n* regions. With silicon diodes, it's typically less than 1 Ω.

conventional current Current in the same direction as holes. The diode arrow points in the forward direction for conventional current.

depletion layer A narrow region on both sides of the junction where recombination has eliminated the holes and free electrons. This recombination produces ions (charged atoms), which are responsible for the barrier potential.

forward bias The easy direction for current. It occurs when the applied voltage makes the *p* side of the diode more positive than the *n* side.

ideal diode The simplest approximation for a rectifier diode. The circuit equivalent is a switch that is closed when forward-biased and open when reverse-biased.

knee voltage Approximately equal to barrier potential. It is the forward voltage beyond which diode current rapidly increases.

rectifier diode The most common type of diode. A rectifier diode is optimized for its unidirectional conduction. It is equivalent to a one-way switch

26

that allows large current in one direction and almost no current in the other.

reverse bias The hard direction for current. It occurs when the *n* side of a diode is more positive than the *p* side.

zener effect At higher reverse voltages, the electric field across the junction can be intense enough to pull electrons out of valence orbits. These liberated electrons then produce a large reverse current.

SELF-TESTING REVIEW

Read each of the following and provide the missing words. Answers appear at the beginning of the next question.

1. A *pn* diode is a crystal with *p* material on one side of the junction and *n* material on the other side. After the junction is formed, _____ carriers diffuse across the junction and _____.

2. *(majority, recombine)* This recombination produces the _____ layer, which contains immobile charged atoms. For germanium diodes, the barrier potential is about 0.3 V, and for silicon diodes it is around _____ V.

3. *(depletion, 0.7)* When the applied voltage overcomes the _____ potential, diode current is large. This is known as _____ bias.

4. *(barrier, forward)* When the *n* side is more positive than the *p* side, the diode is _____ biased and only a small current exists.

5. *(reverse)* The diode curve indicates a _____ current exists when the diode voltage is greater than the knee voltage. In the reverse direction only a small current exists unless the diode voltage exceeds the _____ voltage.

6. *(large, breakdown)* The ideal diode is the simplest approximation of a rectifier diode. There is no breakdown, no _____ current, and no forward _____ drop. The equivalent circuit is a switch.

7. *(reverse, voltage)* The second and third approximations take into account the _____ voltage drop across a diode. The second approximation uses _____ V for germanium diodes and _____ V for silicon diodes. The third approximation includes an additional voltage drop caused by _____ resistance.

8. *(forward, 0.3, 0.7, bulk)* A diode has reverse resistance. Silicon diodes

Transistor Circuit Approximations

have a much larger reverse resistance than germanium diodes. This is one reason for the overwhelming use of silicon diodes.

PROBLEMS

2-1. Use the ideal-diode approximation to calculate the current through the diode in Fig. 2-17a.

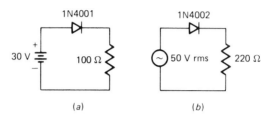

(a)   (b)

Figure 2-17

2-2. In Fig. 2-17b, a sinusoidal source of 50 V rms drives the circuit. Treat the diode as ideal. What is the maximum forward current? The peak inverse voltage?

2-3. Use the second approximation to find the maximum current in Fig. 2-17a and b.

2-4. In Fig. 2-18a, use the ideal-diode approximation to find the current through the diode.

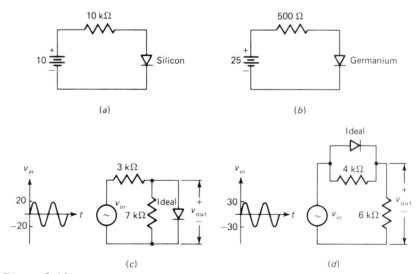

(a)   (b)

(c)   (d)

Figure 2-18

2-5. Calculate the current through the diode in Fig. 2-18*b* using the ideal-diode approach.

2-6. In Fig. 2-18*c*, what is the maximum current through the diode? The peak inverse voltage?

2-7. What is the positive peak output voltage in Fig. 2-18*c*? The negative peak output voltage?

2-8. In Fig. 2-18*d*, what is the maximum current through the diode? The positive peak output voltage? The negative peak output voltage?

2-9. Use the second approximation to calculate the current in Fig. 2-18*a*.

2-10. What is the current in Fig. 2-18*b* using the second approximation?

2-11. In the worst case, a 1N4002 has a reverse current of 50 μA for a reverse voltage of 100 V and a temperature of 100°C. In Fig. 2-17*b*, what is the maximum possible reverse current?

Chapter 3

Special Diodes

Chapter 2 discussed the rectifier diode, the most common type in the electronics industry. The main application of rectifier diodes is to *power supplies* (Chap. 4), in circuits that convert alternating current to direct current. It's possible to optimize other characteristics of the semiconductor diode to produce *special diodes* that are useful in certain applications. This chapter is about these special diodes.

3-1 LIGHT-EMITTING DIODES

In a forward-biased diode, free electrons cross the junction and fall into holes. When they recombine, these electrons radiate energy. In a rectifier diode, the energy goes off as heat. But in a *light-emitting diode* (LED), the energy radiates as light.

By using elements such as gallium, arsenic, and phosphorus, a manufacturer can produce LEDs that radiate red, green, yellow, orange, and infrared (invisible). LEDs that produce visible radiation are used in instrument displays, calculators, digital watches, etc. The infrared LED finds applications in burglar-alarm systems and other areas requiring invisible radiation.

The advantages of an LED over an incandescent lamp are long life (more than 20 years), low voltage (1 to 2 V), and fast on-off switching (nanoseconds).

EXAMPLE 3-1

Figure 3-1*a* shows the schematic symbol for an LED. (The arrows remind us of the emitted light.) If the voltage across the LED is 1.6 V, how much current is there through the LED?

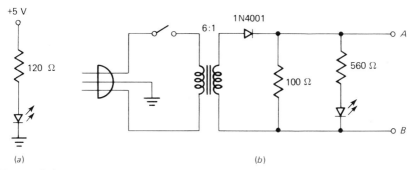

Figure 3-1 LED indicator.

SOLUTION

The voltage across the resistor equals the supply voltage minus the LED voltage. Therefore, the LED current is

$$I = \frac{5\text{ V} - 1.6\text{ V}}{120\ \Omega} = \frac{3.4\text{ V}}{120\ \Omega} = 28.3\text{ mA}$$

EXAMPLE 3-2

When the on-off switch in the primary of Fig. 3-1*b* is closed, 120 V rms drives the primary winding. What is the peak LED current for an LED voltage of 2.1 V?

SOLUTION

The transformer has a step-down ratio of 6:1. Therefore, the secondary voltage is

$$V_2 = \frac{N_2}{N_1} V_1 = \frac{1}{6}\, 120 = 20\text{ V rms}$$

The peak secondary voltage is

$$V_{2(\text{peak})} = \sqrt{2} \times 20\text{ V} = 28.3\text{ V}$$

Treating the 1N4001 as ideal gives an LED current of

$$I = \frac{28.3\text{ V} - 2.1\text{ V}}{560\ \Omega} = \frac{26.2\text{ V}}{560\ \Omega} = 46.8\text{ mA}$$

Transistor Circuit Approximations

EXAMPLE 3-3

Figure 3-2*a* shows a *seven-segment indicator,* an array of LEDs that can display any numeral from 0 through 9. Which segments do we have to activate to display a 7?

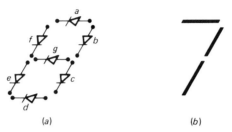

(a)   (b)

Figure 3-2 Seven-segment indicator.

SOLUTION

Easy. Turn on segments *a, b,* and *c.* Because the LEDs have rectangular shapes, you will see the display of Fig. 3-2*b.*

3-2 PHOTODIODES, ISOLATORS, AND SOLAR CELLS

A reverse-biased diode has a small current because of its thermally-produced minority carriers. Experiments have shown that light as well as heat can produce these minority carriers. When the diode is in a glass package, strong light hitting the junction changes the reverse current.

Photodiodes

A *photodiode* is optimized for its sensitivity to light. In this diode, a glass window lets incoming light pass through the package to hit the depletion layer. The incoming light produces holes and free electrons. Figure 3-3*a*

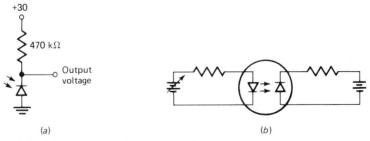

(a)   (b)

Figure 3-3 *(a)* Photodiode. *(b)* Optical isolator.

32

shows the schematic symbol of a photodiode; the arrows represent the incoming light. Especially important, reverse bias is the normal way to operate a photodiode because the incoming light controls the reverse current. The stronger the light is in Fig. 3-3*a*, the greater the voltage drop across the resistor and the lower the output voltage.

Optical isolators

An *optical isolator* combines an LED and a photodiode (or other photodetector) in a single package as shown in Fig. 3-3*b*. The LED supply voltage forces current I_{in} through the LED. The light from the LED hits the photodiode and sets up current I_{out}. If we vary the LED supply, I_{in} varies and this produces a change in I_{out}. In fact, if the LED current has an ac variation, the photodiode current will have an ac variation of the same frequency.

The advantage of an optical isolator is the electrical isolation between the LED circuit and the photodiode circuit; typically, the resistance between the input and output is greater than 10^{10} Ω. This is why the device is called an isolator. The only contact between input and output is the light.

Solar cells

In a *solar cell,* incoming light creates holes and free electrons in the depletion layer. But the similarity between a solar cell and a photodiode ends here. Normally, a solar cell is unbiased because it is used as an energy source. The incoming light produces electron-hole pairs, which produce a voltage across the unbiased junction. This voltage is then used as a dc supply for other circuits. The solar cell is physically larger and more heavily doped than a photodiode; typically, solar-cell current is in milliamperes compared to photodiode current in microamperes.

3-3 SCHOTTKY DIODES

At lower frequencies the *pn* rectifier diode can easily turn off when the bias changes from forward to reverse. But as the frequency increases, we reach a point where the ordinary *pn* diode cannot turn off fast enough to prevent significant reverse current. This section tells you why and discusses how a *Schottky diode* cures the high-frequency problem.

Charge storage

In Fig. 3-4*a*, an ordinary *pn* diode works fine as a rectifier, provided the frequency is not too high. As shown in Fig. 3-4*b*, the output at low frequencies

33

Transistor Circuit Approximations

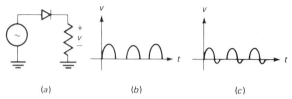

Figure 3-4 *(a)* Half-wave rectifier. *(b)* Normal output. *(c)* Small reverse conduction.

is the half-wave signal discussed in the preceding chapter. Somewhere above 1 MHz, however, the on-off switching action deteriorates and we get some reverse current during the early part of the negative half cycle (see Fig. 3-4c). As the frequency increases, the amount of reverse conduction gets worse until finally the rectifier is useless.

What causes reverse conduction at high frequencies? In a forward-biased diode, free electrons cross the junction, and after a brief interval called the *lifetime,* these free electrons recombine with holes. If the lifetime equals 1 μs, free electrons exist for an average of 1 μs before recombination takes place. In other words, free electrons are temporarily stored in the *p* material for about 1 μs before they fall into holes. This phenomenon is called *charge storage.*

If you suddenly reverse-bias the diode, the stored charges can flow in the reverse direction for about 1 μs. At lower frequencies the reverse flow of stored charges is insignificant. But as the frequency increases, this reverse flow is what accounts for the reverse conduction shown in Fig. 3-4c. The greater the lifetime, the longer the stored charges can contribute to reverse current.

The Schottky diode

The Schottky diode uses a metal such as gold, silver, or platinum on one side of the junction and doped silicon (usually *n*-type) on the other side (see Fig. 3-5a). When the Schottky diode is unbiased, free electrons on the *n* side are in smaller orbits than the free electrons in the metal. This difference in orbit size is called the *Schottky barrier.*

When the diode is forward-biased, free electrons on the *n* side gain enough energy to travel in larger orbits. Because of this, free electrons can cross the junction and enter the metal, producing a large current. Because the metal has no holes, there is no depletion layer and no charge storage (approximately zero lifetime).

The lack of charge storage means the Schottky diode can switch off faster than an ordinary *pn* rectifier diode. In fact, a Schottky diode easily rectifies

34

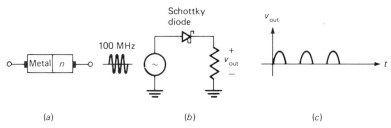

(a)   (b)   (c)

Figure 3-5 Schottky diode. *(a)* Structure. *(b)* Half-wave rectifier. *(c)* Rectified output.

frequencies above 300 MHz. For instance, Fig. 3-5*b* shows a half-wave recti-fier using a Schottky diode (notice the schematic symbol). Even though the input frequency is 100 MHz, the output is almost a perfect half-wave signal (Fig. 3-5*c*).

The most important application of Schottky diodes is in digital electronics. The speed of computers is limited by how fast their switching devices can turn on and off. This is where the Schottky diode comes in. Because it has no charge storage, the Schottky diode has become the backbone of an important family of digital integrated circuits known as *Schottky TTL* (transis-tor-transistor logic).

3-4 VARACTORS

When reverse-biased, a diode has a large resistance R_R. This resistance de-pends on the minority carriers and surface leakage. Besides this resistance, a diode also has a built-in capacitance called the *transition capacitance*.

Transition capacitance

In Fig. 3-6*a*, the depletion layer acts like an insulator. Because the *p* and *n* regions are fairly good conductors, we can visualize a parallel-plate capacitor: the depletion layer is the dielectric, and the *p* and *n* regions are the plates of the capacitor.

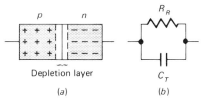

(a)   (b)

Figure 3-6 Junction capacitance.

35

Transistor Circuit Approximations

Figure 3-6*b* is the equivalent circuit for a reverse-biased diode. A large resistance R_R is in parallel with a transition capacitance C_T. The value of transition capacitance (also known as barrier capacitance, junction capacitance, or depletion-layer capacitance) decreases when the reverse voltage increases. Why? Because the depletion layer gets wider for larger reverse voltage, equivalent to moving the parallel plates apart. In other words, a reverse-biased diode acts like a variable capacitor.

The varactor

Silicon diodes optimized for their variable capacitance are called *varactors*. Figure 3-7*a* shows the schematic symbol for a varactor. Because it acts like a variable capacitance, the varactor is replacing the mechanically tuned capacitor in many applications. For instance, an inductor in parallel with a capacitor has a resonant frequency of

$$f = \frac{1}{2\pi \sqrt{LC}} \qquad \textbf{(3-1)}$$

If we use a varactor for the capacitor, we can change the resonant frequency by varying the reverse voltage rather than by turning the shaft of a movable-plate capacitor.

Figure 3-7*b* illustrates this idea of electronic tuning. Notice the varactor is reverse-biased. As the tuning control is varied, the reverse voltage changes. This changes the capacitance of the varactor and affects the rest of the circuit. If the varactor is part of a resonant circuit, the resonant frequency changes. This is how we can tune in different radio or television stations.

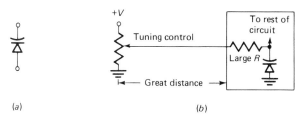

(a)   (b)

Figure 3-7 *(a)* Varactor symbol. *(b)* Remote tuning.

3-5 ZENER DIODES

The breakdown region of a *pn* diode can be made very sharp and almost vertical. Diodes with an almost vertical breakdown region are known as *zener diodes*.

Avalanche and zener effects

Breakdown is caused by either of two effects: avalanche or zener. When a diode is reverse-biased, minority carriers flow in the reverse direction. For higher reverse voltages, these minority carriers can reach sufficient velocities to knock valence electrons out of their shells. These released electrons become free electrons and can attain sufficient velocity to dislodge more valence electrons. The resulting avalanche of free electrons produces a large reverse current.

Zener effect is different. The electric field across the junction can become intense enough to pull valence electrons directly out of their shells. This produces a large reverse current. Zener effect is sometimes called *high-field emission* because it is the electric field that produces free electrons.

When a diode breaks down, either avalanche or zener effect predominates. Below 6 V, the zener effect is more important. Above 6 V, the avalanche effect takes over. Diodes with breakdown voltages greater than 6 V should be called avalanche diodes. But the general practice in industry is to refer to diodes exhibiting either effect as *zener diodes.*

Equivalent circuits

Figure 3-8a shows the curve of a zener diode. In the forward direction, conduction takes place at approximately 0.7 V. In the reverse direction, breakdown occurs at the zener voltage V_Z. The current in the breakdown region depends on the external circuit connected to the diode. As long as the external resistance limits the zener current to a safe value, the zener diode is not destroyed.

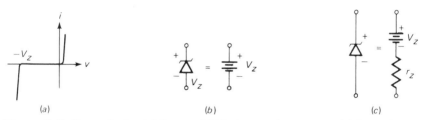

Figure 3-8 Zener diode. *(a)* Curve. *(b)* Ideal equivalent circuit. *(c)* Second approximation.

To analyze zener-diode circuits quickly and easily, we can use the ideal approximation shown in Fig. 3-8b. A zener diode (note the schematic symbol) operating in the breakdown region is equivalent to a battery. Because of this, the current through the zener diode can change but the voltage will

37

Transistor Circuit Approximations

remain constant. It is this constant voltage that has made the zener diode an important device in *voltage regulation* (holding voltage constant).

The breakdown region of Fig. 3-8a is not quite vertical. To account for the slight increase in reverse voltage as current increases, we can include a small resistance as shown in Fig. 3-8c. This resistance is called the *zener resistance* (also known as the zener impedance). The battery-resistor combination of Fig. 3-8c is our second approximation for a zener diode.

EXAMPLE 3-4

The 1N758 of Fig. 3-9a has a breakdown voltage of 10 V and a zener resistance of 17 Ω. If V_{IN} can vary from 20 to 40 V, what are the minimum and maximum zener current? The minimum and maximum output voltage?

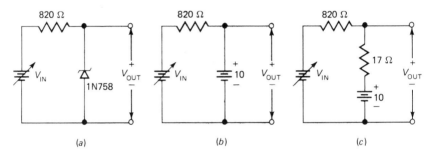

(a)   (b)   (c)

Figure 3-9

SOLUTION

First, let us solve the problem using the ideal approximation, shown in Fig. 3-9b. The applied voltage (20 to 40 V) is always greater than the breakdown voltage. This is why the zener diode ideally acts like a battery of 10 V. Therefore, ideally the output voltage is constant with a value of

$$V_{OUT} = 10 \text{ V}$$

for any V_{IN} between 20 and 40 V.

The minimum zener current occurs for minimum source voltage. With Ohm's law applied to the resistor,

$$I_{Z(min)} = \frac{20 \text{ V} - 10 \text{ V}}{820 \ \Omega} = 12.2 \text{ mA}$$

The maximum zener current exists when the source voltage is maximum:

38

$$I_{Z\,(\text{max})} = \frac{40\text{ V} - 10\text{ V}}{820\ \Omega} = 36.6\text{ mA}$$

Using the second approximation changes the answers slightly. Figure 3-9c shows the zener diode replaced by a battery and resistor. The minimum zener current is

$$I_{Z\,(\text{min})} = \frac{20\text{ V} - 10\text{ V}}{837\ \Omega} = 11.9\text{ mA}$$

and the maximum zener current is

$$I_{Z\,(\text{max})} = \frac{40\text{ V} - 10\text{ V}}{837\ \Omega} = 35.8\text{ mA}$$

The output voltage is slightly more than 10 V because of the drop across the zener resistance. The minimum output voltage is

$$V_{\text{OUT(min)}} = 10 + (0.0119 \times 17) = 10.2\text{ V}$$

and the maximum output is

$$V_{\text{OUT(max)}} = 10 + (0.0358 \times 17) = 10.6\text{ V}$$

The point of this example is to illustrate *voltage regulation*. Here we have a source that varies from 20 to 40 V, a 100-percent change. To a second approximation, the output voltage varies from 10.2 to 10.6 V, a 3.9-percent change. Therefore, the zener diode has reduced a 100-percent change in input to only a 3.9-percent change in output.

Chapter 4 discusses the effect of a load resistance connected across the zener diode. As will be shown, changes in load resistance have almost no effect on the load voltage. In other words, a zener diode holds the output voltage approximately constant in spite of changes in source voltage and load resistance.

3-6 LASER DIODES

In a forward-biased LED, free electrons cross the junction and recombine with holes. As each free electron falls from a larger orbit to a smaller one, it loses energy. Where does the energy go? Into the creation of a *photon* (the smallest unit of light). The newly created photon has energy E exactly equal to the energy lost by the falling electron. The color of the emitted light depends on the energy of the photon.

The light from an LED is *monochromatic* (one color) because each electron falls through the same distance when recombining with a hole. But the photons of this light have different *phases,* because they are created at random points in time. Therefore, even though the emitted light is monochromatic, the waves of this light have every possible phase angle between 0° and 360°.

A *laser diode* is different. In this diode, opposite ends of the junction are polished to get mirrorlike surfaces. When free electrons fall into holes, the emitted photons reflect back and forth between the mirror surfaces. As a result, two things happen. First, the region between the mirrored ends acts like a *cavity* (microwave-resonant tank) that filters the light and purifies its color. Second, as the photons bounce back and forth, they induce an avalanche effect that causes all newly created photons to be emitted with the *same phase.*

One of the mirror surfaces is semitransparent; this allows some of the laser light to escape as a fine, threadlike beam of photons (see Fig. 3-10a). This escaping light is like nothing found in nature because it is monochromatic and *coherent* (one phase). In other words, almost all the photons of laser light have the same frequency and phase. This is much different from ordinary sunlight, where the photons have frequencies over a $2:1$ range and phases from 0 to 360°.

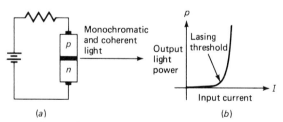

Figure 3-10 Laser diode. *(a)* Circuit. *(b)* Curve.

The output light power versus current for a typical laser diode looks like Fig. 3-10b. The knee of this curve is called the *lasing threshold.* Below this knee, the diode puts out noncoherent light similar to an LED. But above the knee, photon avalanche occurs and the light becomes coherent.

3-7 STEP-RECOVERY DIODES

A *step-recovery diode* takes advantage of charge storage. To begin with, it has a doping profile like Fig. 3-11a. This profile tells us that the concentration

40

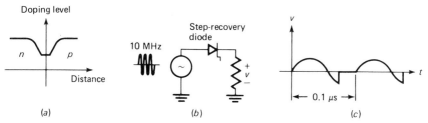

Figure 3-11 Step-recovery diode. *(a)* Doping profile. *(b)* Circuit. *(c)* Rectified output.

of impurities drops off near the junction. An advanced mathematical derivation proves the following about such a doping profile: When the diode goes from forward to reverse bias, free electrons and holes temporarily move away from the junction, resulting in a reverse current. This reverse current comes to an abrupt halt at some point during the negative half cycle.

As an example, Fig. 3-11*b* shows a 10-MHz source driving a step-recovery diode (notice the unusual schematic symbol). During the forward half cycle, the diode conducts the same as any diode. During the reverse half cycle, we get reverse current while the depletion layer is adjusting; then, all of a sudden, the current drops to zero. Figure 3-11*c* shows the output voltage. It's as though the diode has suddenly snapped open during the reverse half cycle; this is why the device is sometimes called a *snap diode.*

Step-recovery diodes are used in pulse and digital circuits for generating very fast pulses. The sudden snap-off of reverse current can produce on-off switching times of less than 1 ns. The step-recovery diode is also used in *frequency multipliers,* circuits that generate harmonics of the input frequency; by filtering the waveform of Fig. 3-11*c,* we can produce a sinusoid whose frequency is a multiple of the input frequency.

3-8 BACKWARD DIODES

Zener diodes normally have breakdown voltages greater than 2 V. By increasing the doping level, however, we can get the zener effect below 2 V. In fact, we can get breakdown near zero as shown in Fig. 3-12*a.* Forward conduction still occurs around 0.7 V, but now reverse conduction (breakdown) starts at 0.1 V or thereabouts.

A diode with a curve like Fig. 3-12*a* is called a *backward diode* because it conducts better in the reverse than in the forward direction. As an example, Fig. 3-12*b* shows a sine wave with a peak of 0.5 V driving a backward diode. (Notice the zener-diode symbol is used for the backward diode.) The 0.5 V is not enough to forward-bias the diode into conduction, but it

41

Transistor Circuit Approximations

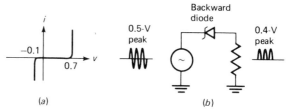

Figure 3-12 Backward diode. *(a)* Curve. *(b)* Low-level rectifier.

is enough to break down the diode. For this reason, the output is a half-wave signal with a peak of 0.4 V (0.1 V is lost across the diode).

Backward diodes are sometimes used to rectify weak signals whose peak amplitudes are between 0.1 and 0.7 V.

3-9 TUNNEL DIODES

By increasing the doping level of a backward diode, we can get breakdown to occur at 0 V. Furthermore, the heavy doping alters the forward curve as shown in Fig. 3-13a. A diode like this is known as a *tunnel diode* (also called an *Esaki diode*).

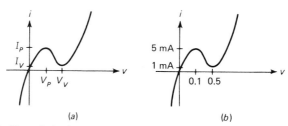

Figure 3-13 Tunnel diode. *(a)* Curve. *(b)* Typical values.

In Fig. 3-13a, forward bias produces immediate conduction. The current reaches a maximum value I_p (peak current) when the diode voltage equals V_p. Then the current decreases to a minimum value I_v (valley current). For diode voltage greater than V_v, current again increases. As an example, Fig. 3-13b shows typical values. The peak current is 5 mA and the valley current is 1 mA.

The region between the peak and valley points is called a *negative-resistance* region. When operating along this part of the curve, a tunnel diode looks like a negative resistance to an ac signal, because an increase in voltage produces a decrease in current. A negative resistance means the diode *produces* ac power instead of absorbing it.

42

For instance, Fig. 3-14 shows a dc source driving a tunnel diode (notice the new symbol). Ordinarily, we don't expect an ac signal out of a dc circuit. Nevertheless, the circuit of Fig. 3-14 can generate an ac output signal. When biased in the negative-resistance region, the tunnel diode converts dc power from the battery into ac power. The frequency of the ac signal equals the resonant frequency of the *LC* tank circuit. The circuit of Fig. 3-14 is called a tunnel-diode oscillator; it is used to generate very high frequencies.

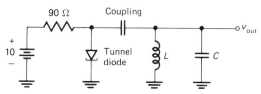

Figure 3-14 Tunnel-diode oscillator.

GLOSSARY

charge storage In a forward-biased diode, charges are temporarily stored near the junction for a brief period known as the lifetime. If the biasing suddenly reverses, these stored charges can flow in the reverse direction for the remainder of their lifetime.

laser diode A diode that emits monochromatic and coherent light above its lasing threshold.

light-emitting diode Better known as an LED, this diode emits monochromatic light when forward-biased.

optical isolator An LED and a photodetector in a single package.

photodiode External light falling on this kind of diode controls the reverse current. The greater the light, the greater the current.

Schottky diode Also called a hot-carrier diode, it can switch on and off much faster than a *pn* diode. The Schottky diode has metal on one side and doped silicon on the other.

varactor A *pn* diode optimized for its variable capacitance when reverse-biased.

voltage regulator A device or circuit that converts a varying input voltage into a constant output voltage.

zener diode A diode designed to operate in the breakdown region. It is used for voltage regulation.

SELF-TESTING REVIEW

Read each of the following and provide the missing words. Answers appear at the beginning of the next question.

1. In a forward-biased diode, free electrons cross the junction and fall into _____. When they recombine, these electrons radiate energy. In an LED, this energy radiates as _____ instead of heat. A seven-segment indicator is an array of LEDs that can display any numeral between 0 and 9.

2. *(holes, light)* A photodiode is optimized for its sensitivity to light. The incoming light produces _____ and free _____ in the depletion layer. The stronger the light, the larger the reverse _____.

3. *(holes, electrons, current)* An optical isolator combines an LED and a photodetector in one package. The advantage of an optical isolator is the electrical isolation between the _____ and _____ circuits.

4. *(input, output)* The lack of charge _____ means the Schottky diode can switch off faster than an ordinary *pn* diode. The most important application of the Schottky diode is in digital electronics; it is the backbone of Schottky TTL.

5. *(storage)* In a reverse-biased diode, the depletion layer acts like an insulator, and the doped regions like capacitor plates. The greater the reverse voltage, the smaller the capacitance. Diodes optimized for this variable capacitance are called _____. They are replacing mechanically tuned capacitors in many applications.

6. *(varactors)* A _____ diode operating in the breakdown region is ideally equivalent to a _____. In the second approximation, we include a small resistance. The main use for this kind of diode is in voltage regulation, holding voltage constant.

7. *(zener, battery)* A _____ diode produces light that is monochromatic and coherent. A step-recovery diode suddenly turns off during the reverse half cycle; it is used in pulse and digital circuits, and in frequency multipliers. Backward diodes conduct better in the _____ direction.

8. *(laser, reverse)* The region between the peak and valley points of a tunnel diode is called a negative-resistance region. When operating along this part of the curve, the tunnel diode converts dc input power to ac output power.

PROBLEMS

3-1. The LED of Fig. 3-15a has 1.8 V across it. What is the current?

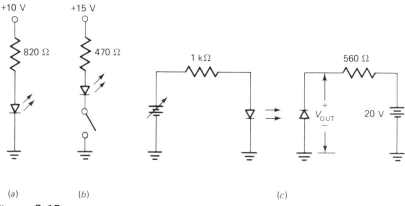

(a)   (b)   (c)

Figure 3-15

3-2. In Fig. 3-15b, the LED has 2 V across it when the switch is closed. What is the LED current when the switch is closed?

3-3. The LED of Fig. 3-15c has a drop of 2.1 V when the source is 5 V. The resulting photodiode current is 10 μA.

a. How much LED current is there?
b. What does V_{OUT} equal?

3-4. When the source of Fig. 3-15c equals 10 V, the LED has 2.3 V across it and the photodiode current is 25 μA. What is the LED current? The value of V_{OUT}?

3-5. The input signal of Fig. 3-16a has a peak of 10 V. The Schottky

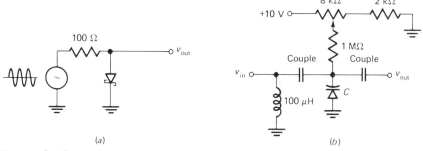

(a)   (b)

Figure 3-16

45

Transistor Circuit Approximations

diode has a knee voltage of 0.4 V. Ideally, what does the output voltage waveform look like? To a second approximation, what does it look like?

3-6. In Fig. 3-16*b*, what is the minimum reverse voltage across the varactor? The maximum?

3-7. The varactor of Fig. 3-16*b* is in parallel with the 100 μH because the capacitors labeled "Couple" appear as ac shorts. If the varactor's capacitance can change from 10 to 40 pF, what is the minimum resonant frequency? The maximum?

3-8. In Fig. 3-17*a*, what is the minimum and maximum current through the zener diode? (Use the ideal approximation.)

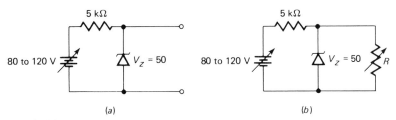

(a)   (b)

Figure 3-17

3-9. If the zener resistance is 125 Ω in Fig. 3-17*a*, what is the minimum and maximum voltage across the zener diode?

3-10. The zener diode of Fig. 3-17*a* has a tolerance of ± 5 percent. What is the minimum and maximum current for a source voltage of 100 V? (Ignore r_Z.)

3-11. In Fig. 3-17*b*, $R = 20$ kΩ. Find the minimum and maximum current through the 5 kΩ. (Use the ideal approximation.)

3-12. If $R = 15$ kΩ in Fig. 3-17*b*, what is the minimum zener current? The maximum?

3-13. In Fig. 3-17*b*, find the value of R that causes the zener diode to come out of the breakdown region for the following source voltages: *(a)* 80 V; *(b)* 120 V.

3-14. If the zener diode of Fig. 3-17*b* has a tolerance of ± 10 percent, what is the minimum possible zener current when $R = 20$ kΩ?

46

Chapter 4

Diode Applications

This chapter begins with diode rectifiers commonly found in power supplies. After discussing the capacitor-input filter, we get into voltage multipliers. Next, we take up zener regulators, clippers, and clampers. The chapter ends with OR gates and AND gates, two logic circuits used in computers.

4-1 THE HALF-WAVE RECTIFIER

Power companies in the United States supply a nominal line voltage of 120 V rms at 60 Hz. (In Europe it is 240 V at 50 Hz.) To get dc voltage from this ac line voltage, we can use a half-wave rectifier like that in Fig. 4-1a. When connected to a power outlet, the three-wire plug grounds the chassis (middle prong) and delivers 120 V rms to the circuit. Figure 4-1b shows how this line voltage looks. As you know from basic circuit theory, the relationship between rms and peak values is

$$V_{RMS} = \frac{V_P}{\sqrt{2}} \qquad \textbf{(4-1)}$$

Average value and output frequency

As discussed in earlier chapters, the half-wave rectifier produces a half-wave output as shown in Fig. 4-1c. The *average* value, also known as the dc value, of this half-wave signal is

$$V_{\text{DC}} = \frac{V_P}{\pi} \qquad \textbf{(4-2)}$$

Transistor Circuit Approximations

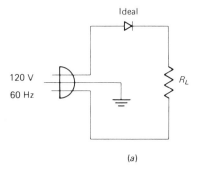

(a)

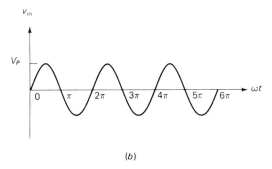

(b)

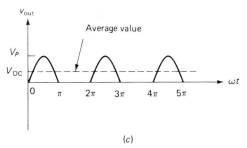

(c)

Figure 4-1 Half-wave rectifier. *(a)* Circuit. *(b)* Input. *(c)* Output.

where $\pi \cong 3.14$. For example, if the peak value is 170 V, the average value of the half-wave output is

$$V_{DC} = \frac{170 \text{ V}}{\pi} = 54.1 \text{ V}$$

The period of the output signal is the same as the input. In other words, with a half-wave rectifier the output frequency equals the input frequency. Therefore, if the sine-wave input has a frequency of 60 Hz, the half-wave output has a frequency of 60 Hz.

48

Transformer and peak inverse voltage

Most electronics equipment is transformer-coupled at the input, as shown in Fig. 4-2a. The transformer allows us to step the voltage up or down, as needed. Another advantage is isolation from the power line; this reduces the risk of dangerous electrical shock.

Figure 4-2b shows the half-wave rectifier at the peak of the negative half cycle of secondary voltage. The diode is shaded or dark to indicate that it is off. Since there is no current through the diode, the maximum secondary

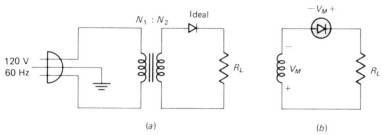

(a)   (b)

Figure 4-2 Transformer-coupled half-wave rectifier.

voltage appears across the diode. As mentioned in Chap. 2, this maximum reverse voltage is called the peak inverse voltage (PIV). In symbols,

$$PIV = V_M \qquad\qquad (4\text{-}3)$$

EXAMPLE 4-1

The turns ratio of Fig. 4-2a is $N_1:N_2 = 4:1$. What is the maximum voltage across the secondary winding? The average voltage across R_L? The PIV across the diode?

SOLUTION

The peak primary voltage is

$$V_P = \sqrt{2} \times V_{RMS} = \sqrt{2} \times 120\ V = 170\ V$$

The maximum secondary voltage is

$$V_M = \frac{N_2}{N_1} V_P = \frac{1}{4} \times 170\ V = 42.5\ V$$

With an ideal diode, the average output voltage is

49

Transistor Circuit Approximations

$$V_{DC} = \frac{V_M}{\pi} = \frac{42.5 \text{ V}}{\pi} = 13.5 \text{ V}$$

The peak inverse voltage is

$$\text{PIV} = V_M = 42.5 \text{ V}$$

4-2 THE CENTER-TAP RECTIFIER

Figure 4-3a shows a center-tap rectifier. During the positive half cycle of secondary voltage, the upper diode is forward-biased and the lower diode is reverse-biased; therefore, current is through the upper diode, the load resistor, and the upper-half winding of Fig. 4-3c. During the negative half cycle, current is through the lower diode, the load resistor, and the lower-half winding of Fig. 4-3d. Especially important, the load current (current through R_L) is in the same direction during both half cycles. This is why the load voltage is the *full-wave* signal shown in Fig. 4-3b.

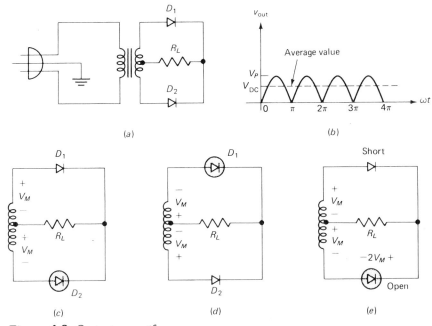

Figure 4-3 Center-tap rectifier.

Average value and output frequency

The average value of a full-wave signal is twice that of a half-wave signal. The formula is

$$V_{DC} = \frac{2V_P}{\pi} \tag{4-4}$$

For example, if the peak value is 170 V, the average value of a full-wave signal is

$$V_{DC} = \frac{2 \times 170 \text{ V}}{\pi} = 108 \text{ V}$$

In Fig. 4-3b, each cycle of input produces two cycles of output. This is why the output frequency of a full-wave signal is twice the input frequency. Given an input frequency of 60 Hz, the output frequency is 120 Hz.

Peak inverse voltage

Figure 4-3e shows the circuit at the instant the secondary voltage reaches its maximum value. V_M is the maximum voltage across half the secondary winding; therefore, the reverse voltage across the nonconducting diode has a peak inverse value of

$$PIV = 2V_M \tag{4-5}$$

EXAMPLE 4-2

If $V_M = 28.3$ V in Fig. 4-3a, what is the average output voltage? The PIV?

SOLUTION

The average load voltage is

$$V_{DC} = \frac{2V_M}{\pi} = \frac{2 \times 28.3 \text{ V}}{\pi} = 18 \text{ V}$$

The peak inverse voltage across the diodes is

$$PIV = 2V_M = 2 \times 28.3 \text{ V} = 56.6 \text{ V}$$

51

4-3 THE BRIDGE RECTIFIER

Figure 4-4a shows a *bridge* rectifier, the most widely used of all. During the positive half cycle of secondary voltage, diodes D_2 and D_3 are forward-biased; therefore, conventional load current is to the left in Fig. 4-4c. During the negative half cycle, diodes D_1 and D_4 are forward-biased, and the conventional load current is to the left in Fig. 4-4d. On either half cycle, the load current is in the same direction. This is why the load voltage is the full-wave signal shown in Fig. 4-4b.

The average load voltage is

$$V_{DC} = \frac{2V_P}{\pi} \qquad \textbf{(4-6)}$$

The output frequency is twice the input:

$$f_{out} = 2f_{in} \qquad \textbf{(4-7)}$$

The peak inverse voltage is

$$\text{PIV} = V_M$$

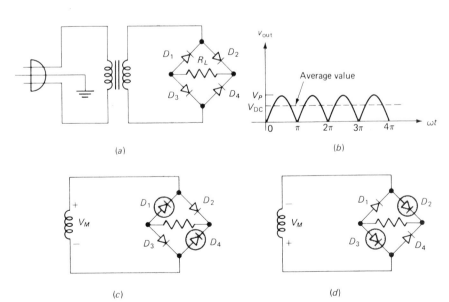

(a)

(b)

(c)

(d)

Figure 4-4 Bridge rectifier.

As mentioned, the bridge rectifier is the most popular rectifier circuit in power supplies. Its main disadvantage is having four diodes, two of which conduct on alternate half cycles. This creates a problem when low secondary voltages are involved, because the two diode drops (1.4 V) become significant. For this reason, the center-tap rectifier may be preferred for low-voltage applications because it has only one diode drop (0.7 V).

EXAMPLE 4-3

The maximum secondary voltage is 68 V in Fig. 4-4a. What are the dc load voltage, output frequency, and peak inverse voltage?

SOLUTION

The average output voltage is

$$V_{DC} = \frac{2V_M}{\pi} = \frac{2 \times 68 \text{ V}}{\pi} = 43.3 \text{ V}$$

The output frequency is twice the input frequency. Since the line frequency is 60 Hz,

$$f_{out} = 2 \times 60 \text{ Hz} = 120 \text{ Hz}$$

The peak inverse voltage is

$$PIV = V_M = 68 \text{ V}$$

4-4 THE CAPACITOR-INPUT FILTER

The half-wave and full-wave signals are pulsating direct voltages. The use for such voltages is limited to charging batteries, running dc motors, and a few other applications. What we really have is a dc voltage that is constant in value, similar to the voltage from a battery. To get a constant voltage like this, we can use a *capacitor-input filter,* the most common filter in rectifier applications.

Basic idea

Figure 4-5a shows a capacitor-input filter. When the diode is on, the source charges the capacitor. Because the diode has low resistance, the charging time constant is very short. On the other hand, when the diode is off, the capacitor discharges through the load. By deliberate design, the discharging

53

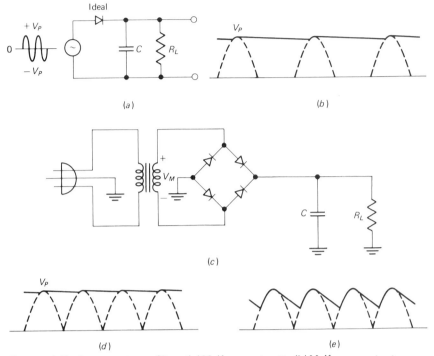

Figure 4-5 Capacitor-input filter. *(a)* Half-wave circuit. *(b)* Half-wave output. *(c)* Full-wave bridge. *(d)* Full-wave output. *(e)* Output with heavy load.

time constant is very long. For this reason, the output voltage looks like Fig. 4-5*b* (solid curve). Notice how the capacitor loses only a small amount of its charge when the diode is off. Then, near the next positive peak, the diode turns on briefly and recharges the capacitor.

The output signal of Fig. 4-5*b* is almost a pure dc voltage equal to the peak input voltage. This is why the circuit is sometimes called a *peak rectifier* or *peak detector.* The only deviation from pure dc is the small *ripple* (fluctuation) caused by charging and discharging the capacitor. The smaller the ripple is, the better the filtering action.

Full-wave action

Figure 4-5*c* shows a bridge rectifier working into a capacitor-input filter. This results in better peak rectification because the capacitor charges twice as often (see Fig. 4-5*d*). As a result, the ripple is cut in half and the dc output more closely approaches the peak voltage. Full-wave peak rectifiers are used far more often than half-wave peak rectifiers. From now on, we will concentrate on the full-wave capacitor-input filter.

Nonideal peak rectification

The most common rectifier circuit is a full-wave bridge driving a capacitor-input filter (Fig. 4-5c). Ideally, the output voltage is a ripple-free dc voltage equal to the maximum secondary voltage V_M, assuming ideal diodes, a zero charging time constant, and an infinite discharging time constant.

Actually, the charging time constant R_SC is not zero, because there is some source resistance R_S in the charging path. R_S includes the resistance of the transformer windings seen from the secondary side; it also includes the bulk resistances of the diodes. Furthermore, the discharging time constant R_LC is not infinite. In fact, when the load current is large, R_L is small and so too is the discharging time constant. This increases the ripple and decreases the dc voltage or average voltage (Fig. 4-5e).

Using a computer, it's possible to analyze the effects of source and load resistance. Figure 4-6 summarizes the results in the form of normalized output curves. The top curve is for zero source resistance. As you can see, when the discharging time constant R_LC increases, the normalized output V_{DC}/V_M approaches unity. This means V_{DC} is approaching the value of V_M, the maximum output voltage.

The second curve is for $R_S/R_L = 0.01$. In this case, source resistance is one percent of load resistance. When the discharging time constant increases, the normalized output V_{DC}/V_M approaches a maximum value of 0.95. This means the dc output voltage can be no more than 95 percent of the maximum voltage driving the capacitor-input filter.

The remaining curves are self-explanatory. In all cases, notice how a short discharging time constant (small R_LC) causes the dc output voltage to decrease. Even when the discharging time constant is very long, too large a source resistance means less-than-maximum dc output. For instance, when $R_S/R_L = 0.08$, V_{DC}/V_M can be no greater than 0.78.

With the curves of Fig. 4-6, you can analyze any full-wave rectifier with a capacitor-input filter for a variety of source and load conditions. Notice that the curves are for a line frequency of 60 Hz.

EXAMPLE 4-4

$R_S = 1\ \Omega$ in Fig. 4-7a. What is the dc output voltage for $V_M = 25$ V?

SOLUTION

The discharging time constant is

Transistor Circuit Approximations

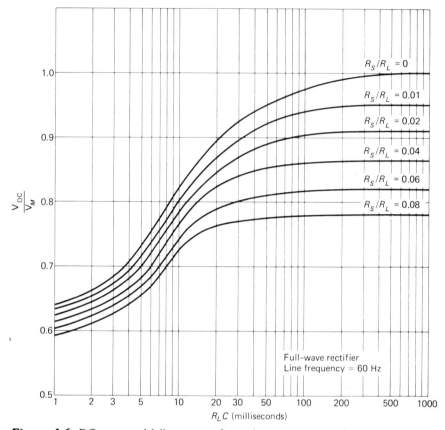

Figure 4-6 DC output of full-wave rectifier with capacitor-input filter.

$$R_L C = 100 \ \Omega \times 1000 \ \mu F = 100 \ \text{ms}$$

Since $R_S/R_L = 1/100 = 0.01$, read the second curve of Fig. 4-6 to get

$$\frac{V_{DC}}{V_M} \cong 0.94$$

or $\qquad\qquad V_{DC} \cong 0.94 V_M = 0.94 \times 25 \ \text{V} = 23.5 \ \text{V}$

EXAMPLE 4-5

Sometimes a *surge resistor* is included to protect the diodes from excessive current when the power is turned on in a circuit like Fig. 4-7b. The capacitor is initially uncharged, and a large current exists for the first few cycles when

56

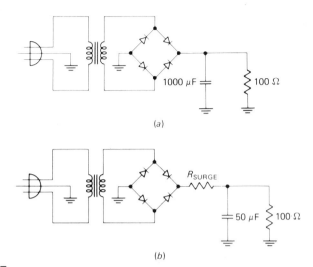

(a)

(b)

Figure 4-7

the power is turned on. The surge resistance is part of the source resistance facing the capacitor; therefore, it limits the initial current.

If $R_S = 4 \ \Omega$, what is the dc output voltage for $V_M = 25$ V?

SOLUTION

This time, $R_S/R_L = 4/100 = 0.04$. The discharging time constant is

$$R_L C = 100 \ \Omega \times 50 \ \mu\text{F} = 5 \text{ ms}$$

The fourth curve of Fig. 4-6 gives

$$\frac{V_{DC}}{V_M} \cong 0.68$$

or $\qquad V_{DC} \cong 0.68 V_M = 0.68 \times 25 \text{ V} = 17 \text{ V}$

4-5 L AND π FILTERS

Figure 4-8a shows a full-wave rectifier driving a *choke* (iron-core inductor), a capacitor, and a load resistor. The full-wave signal out of the rectifier has a dc component and an ac component (see Fig. 4-8b). The choke allows the dc component to pass through easily because X_L is zero for dc or constant current. Since the capacitor appears open at zero frequency, all the dc current out of the choke flows through load resistance R_L. A *choke-input filter* like Fig. 4-8a is often called an *L filter*.

57

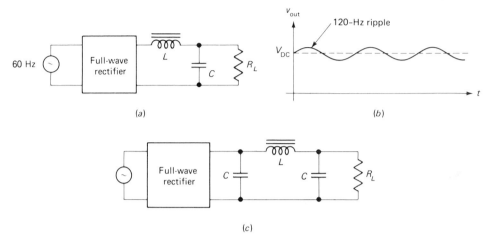

Figure 4-8 *(a)* L filter. *(b)* Output voltage. *(c)* π filter.

Ripple

The ac component out of the rectifier has a frequency of 120 Hz. The choke blocks this ac component because X_L is high at this frequency. Furthermore, any ac current that does manage to get through the choke passes through the capacitor (very low X_C) rather than through R_L. In other words, the choke and capacitor act like an ac voltage divider that *attenuates* (reduces) the ripple.

The choke-input filter is excellent for attenuating ripple, but it's rarely used because chokes are bulky and expensive.

The π filter

Figure 4-8c is a π filter. The rectifier works into a capacitor. This produces peak rectification. Then the L section attenuates any ripple out of the first capacitor. By deliberate design, X_L is much greater than X_C. Because of this, the ripple is greatly attenuated as it passes through the L section. Typically, X_L is at least 10 times greater than X_C; therefore, ripple is reduced by at least a factor of 10.

4-6 VOLTAGE MULTIPLIERS

A *voltage multiplier* is two or more peak rectifiers that produce a dc voltage equal to a multiple of the peak input voltage ($2V_P$, $3V_P$, $4V_P$, and so on). These circuits are used for high voltage/low current applications such as

58

supplying cathode-ray tubes (the picture tubes in TV receivers, oscilloscopes, and computer displays).

Voltage doubler

Figure 4-9a shows a *voltage doubler,* a connection of two peak rectifiers. At the peak of the negative half cycle, D_1 is forward-biased and D_2 is reverse-biased. This charges C_1 to the peak voltage V_P with the polarity shown in Fig. 4-9b. At the peak of the positive half cycle, D_1 is reverse-biased and D_2 is forward-biased. Because the source and C_1 are in series, C_2 will charge toward $2V_P$. After several cycles, the voltage across C_2 equals $2V_P$ as shown in Fig. 4-9c.

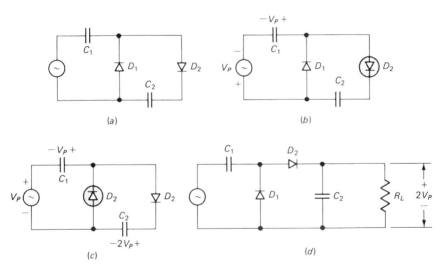

Figure 4-9 Voltage doubler.

By redrawing the circuit and connecting a load resistance, we get Fig. 4-9d. As long as R_L is large, the discharging time constant is long compared to the period, and the output voltage is approximately equal to $2V_P$—that is, provided the load is light (small load current), the dc output voltage is double the peak input voltage.

Voltage tripler

By connecting another section, we get a *voltage tripler* (Fig. 4-10a). The first two peak rectifiers act like a doubler. At the peak of the negative half

Transistor Circuit Approximations

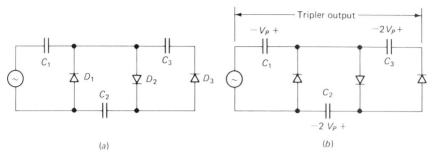

(a)   (b)

Figure 4-10 Voltage tripler.

cycle of input voltage, D_3 is forward-biased. This charges C_3 to $2V_P$ with the polarity shown in Fig. 4-10b. The tripler output appears across C_1 and C_2.

The load resistance is connected across the tripler output. Again, this load resistance must be large to ensure a long discharging time constant. When this condition is satisfied, the dc output equals approximately $3V_P$.

Theoretically, we could continue adding more peak rectifiers to get higher multiples of input peak voltage. But the output ripple keeps getting worse because the discharge between peaks gets larger. For this reason, the doubler and the tripler are the most popular voltage multipliers.

4-7 THE ZENER REGULATOR

Voltage regulation (holding output voltage constant) is poor with a capacitor-input filter because the dc voltage drops off as the load current increases (see Fig. 4-6). Also, if the line voltage changes, the dc output voltage changes. One way to improve the voltage regulation is with a *zener regulator,* shown in Fig. 4-11.

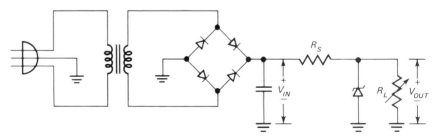

Figure 4-11 Zener regulator.

Basic idea

The voltage from an unregulated power supply such as a peak rectifier is the input voltage V_{IN} to the zener regulator. As long as V_{IN} is greater than V_Z, the zener diode operates in the breakdown region. A series-limiting resistor R_S prevents the zener current from exceeding its rated maximum of I_{ZM}.

Ideally, the zener diode acts like a battery; therefore, the load voltage is constant. For instance, if line voltage increases, V_{IN} will increase. The zener voltage, however, remains approximately constant so that V_{OUT} is approximately equal to V_Z.

Fundamental relationships

In Fig. 4-11, the current through the series-limiting resistor is

$$I_S = \frac{V_{IN} - V_{OUT}}{R_S} \qquad \text{(4-8)}$$

This current splits at the junction of the zener diode and the load resistor. With Kirchhoff's current law,

$$I_Z = I_S - I_L \qquad \text{(4-9)}$$

Ignoring the small zener resistance r_Z, the output voltage is

$$V_{OUT} \cong V_Z \qquad \text{(4-10)}$$

and the load current is

$$I_L = \frac{V_{OUT}}{R_L} \qquad \text{(4-11)}$$

Equations (4-8) through (4-11) are adequate for initial analysis of zener-diode circuits. To improve the analysis, we can include the zener resistance. The change in output voltage for a change in zener current is given by

$$\Delta V_{OUT} = \Delta I_Z r_Z \qquad \text{(4-12)}$$

Maximum limiting resistance

For a zener regulator to hold the output voltage constant, the zener diode must operate in the breakdown region at all times. In other words, there

Transistor Circuit Approximations

must be zener current for all source voltages and load currents. The worst case occurs for minimum source voltage and maximum load current, because at that point the zener current drops to a minimum value.

By solving Eqs. (4-8) through (4-11) simultaneously, we get the maximum allowable series-limiting resistance:

$$R_{S(max)} = \frac{V_{IN(min)} - V_{OUT}}{I_{L(max)}} \tag{4-13}$$

where $R_{S(max)}$ = largest permitted limiting resistor
$\quad V_{IN(min)}$ = smallest possible input voltage
$\quad V_{OUT}$ = zener breakdown voltage
$\quad I_{L(max)}$ = largest possible load current

If you try to use a resistance larger than the value given by this equation, the zener regulator will stop regulating for low source voltage and high load current.

EXAMPLE 4-7

Calculate the current through the series-limiting resistor of Fig. 4-12, the minimum and maximum load current, and the minimum and maximum zener current.

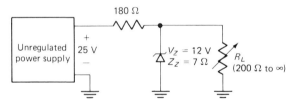

Figure 4-12

SOLUTION

The input voltage is 25 V and the output voltage is 12 V (approximately). Therefore, the current through the 180 Ω is

$$I_S = \frac{V_{IN} - V_{OUT}}{R_S} = \frac{25\text{ V} - 12\text{ V}}{180\ \Omega} \cong 72\text{ mA}$$

The minimum load current occurs when R_L is infinite:

$$I_{L(min)} = 0$$

62

The maximum load current occurs when R_L is 200 Ω:

$$I_{L(max)} = \frac{V_{OUT}}{R_{L(min)}} = \frac{12\text{ V}}{200\text{ }\Omega} = 60\text{ mA}$$

As indicated by Eq. (4-9), the zener current is the difference of the series-limiting current and the load current. Therefore, the minimum zener current is

$$I_{Z(min)} = 72\text{ mA} - 60\text{ mA} = 12\text{ mA}$$

The maximum zener current is

$$I_{Z(max)} = 72\text{ mA} - 0 = 72\text{ mA}$$

Notice these features about the zener regulator. The current through the series-limiting resistor is constant, equal to 72 mA. When the load current increases from 0 to 60 mA, the zener current automatically decreases from 72 to 12 mA. This is normal operation for a zener regulator: any change in load current is compensated for by an equal and opposite change in zener current. In symbols,

$$\Delta I_Z = -\Delta I_L \tag{4-14}$$

Because of this, I_S and V_{OUT} remain fixed in spite of changes in load current.

EXAMPLE 4-8
Calculate the output voltage change in the preceding example by including the zener resistance.

SOLUTION
Ideally, the regulated output voltage is 12 V. To a second approximation, however, changes in zener current produce small changes in output voltage. In the preceding example, the zener current changes from 12 to 72 mA. The zener resistance r_Z (also called the zener impedance Z_Z) is given as 7 Ω in Fig. 4-12. With Eq. (4-12),

$$\Delta V_{OUT} = \Delta I_Z r_Z = (72\text{ mA} - 12\text{ mA}) \times 7\text{ }\Omega = 0.42\text{ V}$$

This tells us the output voltage is not exactly constant; it has a nominal

Transistor Circuit Approximations

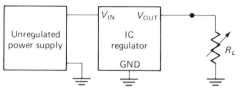

Figure 4-13 IC regulator.

value of 12 V and changes 0.42 V when the load current changes from 0 to 72 mA.

A final point: By combining zener diodes and transistors in an integrated circuit (IC), manufacturers are producing three-terminal voltage regulators (see Fig. 4-13). The input to an IC regulator is the voltage from an unregulated supply such as a peak rectifier. The output of the regulator goes to the load resistance. The new IC regulators are easy to use, offer high reliability, and regulate to better than 1 percent. (This means the change in output voltage is less than 1 percent of the nominal output voltage for all source and load changes.)

EXAMPLE 4-9

A zener regulator has an input voltage from 15 to 20 V and a load current from 20 to 100 mA. To hold load voltage constant under all conditions, what value should the series-limiting resistor have if $V_Z = 10$ V?

SOLUTION

The worst case occurs for minimum source voltage and maximum load current. With Eq. (4-13),

$$R_{S\,(\text{max})} = \frac{15\ \text{V} - 10\ \text{V}}{100\ \text{mA}} = 50\ \Omega$$

If R_S is greater than 50 Ω, the zener regulator will stop regulating for low source voltages and high load currents. In other words, the zener diode will come out of breakdown when $V_{\text{IN}} = 15$ V and $I_L = 100$ mA.

In general, the series-limiting resistor should be as large as possible to protect the zener diode from excessive current. On the other hand, R_S has to be small enough to satisfy Eq. (4-13). Therefore, in this example the correct design value for R_S is 50 Ω.

4-8 CLIPPERS AND CLAMPERS

In radar, computers, and other applications, we sometimes want to remove part of a signal with a *clipper* or shift the dc level with a *clamper.*

64

The clipper

Figure 4-14a shows a positive clipper, a circuit that removes positive parts of an input signal. The circuit works as follows. During the positive half cycle of input voltage, the diode conducts heavily. To a first approximation, it acts like a closed switch. Therefore, the output voltage equals zero during each positive half cycle.

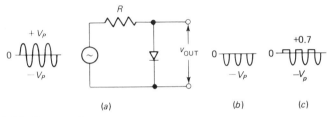

Figure 4-14 Positive clipper.

During the negative half cycle, the diode is reverse-biased and looks like an open switch. As a result, the negative half cycle appears across the output. Ideally, we get the half-wave signal of Fig. 4-14b.

The clipping is not perfect. To a second approximation, a conducting silicon diode drops 0.7 V. Because of this, the output signal is clipped near +0.7 V rather than 0 V (Fig. 4-14c).

By reversing the polarity of the diode, we get a negative clipper, a circuit that removes the negative half cycles.

The clamper

Figure 4-15 is a positive clamper. Ideally, here is how it works. On the first negative half cycle of input voltage, the diode turns on and charges the capacitor to the peak input voltage. Slightly beyond the negative peak, the diode turns off. Because of the long $R_L C$ time constant, the capacitor remains almost fully charged between negative peaks.

According to Kirchhoff's voltage law, the output voltage is the sum of

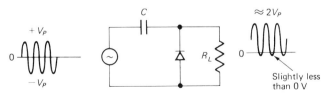

Figure 4-15 Positive clamper.

65

the source voltage and the capacitor voltage. Because of this, the output is a sine wave that is shifted upward as shown in Fig. 4-15. Ideally, the negative peaks of the output voltage fall at 0 V. To a second approximation, these negative peaks are slightly less than 0 V because of the diode drop.

What happens if we reverse the polarity of the diode? The polarity of capacitor voltage reverses, and the circuit becomes a negative clamper. This means the sine wave is shifted downward until the positive peaks fall approximately at 0 V.

Both positive and negative clampers are widely used. Television receivers, for instance, use a clamper to shift the level of the video signal. In television, the clamper is usually called a *dc restorer*.

4-9 LOGIC GATES

Computers use *logic circuits;* these are circuits that duplicate certain mental processes. With logic circuits, we can electronically add, subtract, and process data to solve problems. A *gate* is a logic circuit with one or more input signals but only one output signal. These signals are either low or high voltages.

The OR gate

The OR gate has two or more input signals. If any input signal is high, the output signal is high. Figure 4-16a shows one way to build an OR gate. If both inputs are low, the output is low. If either input is high, the diode with the high input turns on and the output goes high. If both inputs are high, both diodes conduct and the output is high. Because of the two inputs, we call this circuit a 2-input OR gate.

Table 4-1 summarizes the circuit action. Notice that the output is high

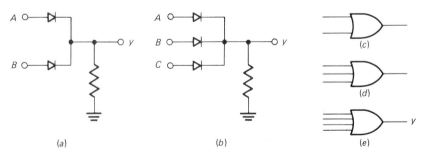

Figure 4-16 OR gate. *(a)* Two-input circuit. *(b)* Three-input circuit. *(c)* Two-input symbol. *(d)* Three-input symbol. *(e)* Four-input symbol.

TABLE 4-1

A	B	y
Low	Low	Low
Low	High	High
High	Low	High
High	High	High

if *A* or *B* is high. This is why the circuit is known as an OR gate. One or more high inputs produce a high output.

Figure 4-16*b* is a 3-input OR gate. If all inputs are low, all diodes are off and the output is low. If one or more inputs are high, the output is high. Table 4-2 summarizes the input-output action. A table like this is called a *truth table;* it lists all the input possibilities and the corresponding outputs.

TABLE 4-2

A	B	C	y
Low	Low	Low	Low
Low	Low	High	High
Low	High	Low	High
Low	High	High	High
High	Low	Low	High
High	Low	High	High
High	High	Low	High
High	High	High	High

An OR gate can have as many inputs as desired; add one diode for each additional input. Six diodes result in a 6-input OR gate, nine diodes in a 9-input OR gate. No matter how many inputs, the action of an OR gate can be summarized like this: One or more high inputs produce a high output.

Figures 4-16*c, d,* and *e* are the symbols used for 2-input, 3-input, and 4-input OR gates. These standard symbols represent OR gates of any design (diode, transistor, etc.).

The AND gate

The AND gate has two or more input signals. All inputs must be high to get a high output. Figure 4-17*a* shows one way to build an AND gate. In this particular circuit the inputs can be either low (ground) or high (+5 V). When both inputs are low (grounded), both diodes conduct and pull

Transistor Circuit Approximations

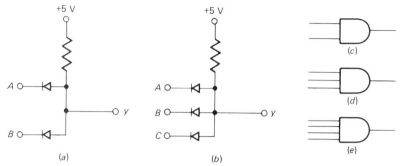

Figure 4-17 AND gate. *(a)* Two-input circuit. *(b)* Three-input circuit. *(c)* Two-input symbol. *(d)* Three-input symbol. *(e)* Four-input symbol.

the output down to a low voltage. If one of the inputs is low and the other high, the diode with the grounded input still conducts and pulls the output down to a low voltage. When both inputs are high, both diodes shut off and the supply voltage pulls the output up to a high voltage (+5 V).

Table 4-3 summarizes the action. As you can see, *both A* and *B* must be high to get a high output; this is why the circuit is called an AND gate.

TABLE 4-3

A	B	y
Low	Low	Low
Low	High	Low
High	Low	Low
High	High	High

Figure 4-17*b* is a 3-input AND gate. If all inputs are low, all diodes conduct and the output is low. Even one conducting diode will pull the output down to a low voltage. Therefore, the only way to get a high output is to have all inputs high. When all inputs are high, none of the diodes is conducting, and the supply voltage pulls the output up to a high voltage. Table 4-4 summarizes the 3-input AND gate.

AND gates can have as many inputs as desired; add one diode for each additional input. Eight diodes, for instance, make an 8-input AND gate; sixteen diodes a 16-input AND gate. No matter how many inputs an AND gate has, the action can be summarized like this: All inputs must be high to get a high output. (The AND gate is sometimes called an *all-or-nothing gate.*)

68

TABLE 4-4

A	B	C	y
Low	Low	Low	Low
Low	Low	High	Low
Low	High	Low	Low
Low	High	High	Low
High	Low	Low	Low
High	Low	High	Low
High	High	Low	Low
High	High	High	High

Figures 4-17c, d, and e are the symbols for 2-input, 3-input, and 4-input AND gates of any design (diode, transistor, etc.).

GLOSSARY

AND gate A logic circuit with two or more inputs and only one output. All inputs must be high to get a high output.

capacitor-input filter The most common type of filter used with rectifiers. In this kind of filter, the first load element facing the rectifiers is a shunt capacitor. Ideally, this capacitor charges to the peak voltage.

clamper In this diode circuit, a series capacitor charges to the peak input voltage. The resulting output is shifted up or down. The original shape of the signal is preserved; only the dc level changes.

clipper A circuit that clips or cuts off certain parts of the input signal.

gate This is a logic circuit with one or more input signals but only one output signal. The output is high only for certain combinations of high input signals.

OR gate In this logic circuit the output is high when one or more inputs are high.

peak inverse voltage Abbreviated PIV, this is the maximum reverse voltage across the diode during the cycle.

peak detector A circuit whose output is a dc voltage equal to the peak voltage of the input signal.

ripple Ideally, the output of a power supply should be a pure dc voltage. Ripple is an unwanted fluctuation in the dc voltage caused by inadequate filtering. When the ripple is small, we ignore it.

Transistor Circuit Approximations

surge resistor When power is first turned on, the diodes of a peak rectifier look into an uncharged capacitor. This uncharged capacitor is equivalent to a short at the first instant. If the Thevenin resistance facing the capacitor is too small, excessive current can destroy the diodes. To prevent this, a resistor can be connected in the charging path of the capacitor to limit the diode current to a safe value. This resistor is called a surge resistor.

voltage multiplier A cascade of peak rectifiers whose dc output voltage equals some multiple of the peak input voltage. A voltage doubler has an output of $2V_p$, a voltage tripler has an output of $3V_p$, and so on.

voltage regulator Any circuit or device that can hold the dc output voltage approximately constant in spite of changes in the supply voltage and load resistance.

SELF-TESTING REVIEW

Read each of the following and provide the missing words. Answers appear at the beginning of the next question.

1. The average value, also known as the _____ value, of a half-wave signal equals V_p/π. The frequency of a half-wave signal is the same as the input frequency.

2. *(dc)* The center-tap rectifier produces a _____ signal. The dc value of this signal is $2V_p/\pi$, and its frequency equals _____ the input frequency.

3. *(full-wave, twice)* The bridge rectifier is the most widely used of all rectifier circuits. It produces a _____ output whose frequency equals twice the input frequency. Because of the two conducting diodes, the diode drops equal 1.4 V with silicon diodes.

4. *(full-wave)* Half-wave and full-wave signals are pulsating direct voltages. The use for such voltages is limited. To get a constant dc voltage, we can use a _____ filter. By deliberate design, the R_LC time constant is much _____ than the period of the input signal. The only deviation from a pure dc output voltage is an ac fluctuation called the _____.

5. *(capacitor-input, longer, ripple)* A voltage multiplier is two or more peak rectifiers that produce a dc voltage equal to a multiple of the peak input voltage. A voltage _____ is a connection of two peak rectifiers

70

that produce a dc output of $2V_P$. A voltage tripler produces an output of $3V_P$.

6. *(doubler)* One way to improve voltage regulation is with a _____ regulator. As long as the input voltage is greater than the breakdown voltage, the zener diode operates in the breakdown region and the output voltage is approximately _____. Ideally, the zener diode acts like a _____ when operating in the breakdown region.

7. *(zener, constant, battery)* In a zener regulator, the current through the series-limiting resistor is constant. When the load current increases, the zener current automatically _____. This is normal, because any change in load current is compensated for by an equal and _____ change in zener current.

8. *(decreases, opposite)* The _____ removes part of the input signal. A _____ preserves the signal shape but changes the dc level.

9. *(clipper, clamper)* Computers use logic circuits that duplicate mental processes. A gate is a logic circuit with one or more inputs but only _____ output. The OR gate has a _____ output when one or more inputs are high. The AND gate has a high output only when _____ inputs are high.

10. *(one, high, all)* A truth table lists all the input/output possibilities.

PROBLEMS

4-1. If the line voltage can be as low as 105 V rms, what is the corresponding peak voltage?

4-2. A half-wave rectifier produces an output with a peak of 45 V. What is the dc value of this output? What is the peak inverse voltage across the diode?

4-3. The output from a center-tap rectifier has a peak of 37 V. What is the average value of the output? The peak inverse voltage across each diode?

4-4. A bridge rectifier produces a full-wave output with a peak of 72 V. What is the dc value of this output? The peak inverse voltage across the diodes?

4-5. In Fig. 4-18, the equivalent load resistance R_L in parallel with the capacitor is 800 Ω. If $R_S = 8$ Ω, what is the dc voltage across the capacitor?

71

Transistor Circuit Approximations

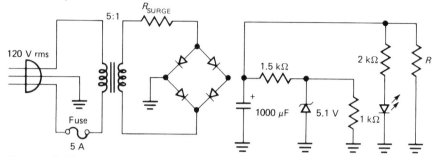

Figure 4-18

4-6. The dc voltage across the capacitor of Fig. 4-18 is 30 V. If the LED has 1.75 V across it, what is the LED current? If the zener diode has a breakdown voltage of 5.1 V, how much current is there through the series-limiting resistor (1.5 kΩ)? How much current is there through the 1 kΩ?

4-7. If R_S (including the surge resistance) is 2 Ω in Fig. 4-18, what is the maximum possible current through the conducting diodes at the instant the on-off switch is closed?

4-8. In Fig. 4-18, the equivalent load resistance R_L seen by the capacitor is 100 Ω. If R_S is zero, what is the $R_L C$ time constant? The dc voltage across the capacitor?

4-9. $R_{SURGE} = 2\ \Omega$ and $R_L = 50\ \Omega$ in Fig. 4-18. Ignore the bulk resistance of the diodes, the primary resistance, and the secondary resistance. What is the dc voltage across the capacitor?

4-10. In Fig. 4-18, $R_S = 0$ and $R_L = 200\ \Omega$.

a. What is the zener current for a line voltage of 120 V rms?
b. If the line voltage falls to 110 V rms, what is the zener current?

4-11. The primary winding resistance is 25 Ω, the secondary winding resistance is 1 Ω, and the bulk resistance of each diode is 1 Ω in Fig. 4-18. If $R_{SURGE} = 4\ \Omega$ and $R_L = 250\ \Omega$, what are each of the following:

a. The dc voltage across the capacitor?
b. The LED current if the LED drop is 2.1 V?
c. The zener current?
d. The current through R if $R = 270\ \Omega$?

4-12. The input to a voltage doubler is 120 V rms. What is the approximate dc output voltage?

72

4-13. The input to a voltage tripler is 360 V rms. What is the ideal dc output?

4-14. The input voltage varies from 12 to 16 V in Fig. 4-19. If the zener

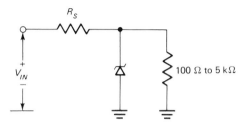

Figure 4-19

diode has a breakdown voltage of 6.2 V, what is the maximum allowable value for R_S?

4-15. In Fig. 4-19, $V_{IN} = 30$ V and $V_Z = 10$ V. What is the largest series-limiting resistance for the range of load resistance shown? What is the minimum zener current? The maximum zener current?

Chapter 5

Small-Signal Diode Approximations

A transistor is like two back-to-back diodes. As mentioned in Chap. 1, the main application of transistors is amplification of weak signals. As a preparation for transistor-circuit analysis, we need to review the *superposition* theorem and to discuss *small-signal* approximations for a diode.

5-1 THE SUPERPOSITION THEOREM

The superposition theorem states the following: In a network with two or more sources, the current or voltage for any component is the algebraic sum of the effects produced by each source acting separately. To have only one active source, all other sources must be reduced to zero; this means replacing voltage sources by short circuits and current sources by open circuits.

An example

Figure 5-1*a* shows an ac source and a dc source, both driving a 10-kΩ resistor. The dc source has a voltage of 10 V, and the ac source has a peak of 10 mV. How can we find the total current through the 10-kΩ resistor using the superposition theorem?

Here's how to get the dc component. Reduce the ac voltage source to zero, identical to replacing it by a short as shown in Fig. 5-1*b*. What remains is the *dc equivalent circuit.* In this simplified circuit the dc current through the resistor is

$$I = \frac{10\,V}{10\,k\Omega} = 1\,mA$$

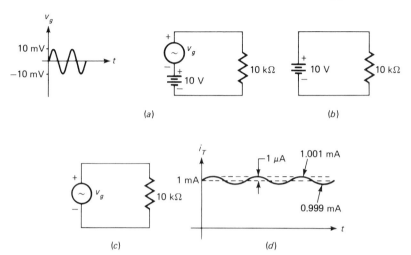

Figure 5-1 Superposition theorem.

Next, find the ac component. Reduce the dc voltage source to zero and what remains is the *ac equivalent circuit* of Fig. 5-1c. The ac source is sinusoidal with a peak of 10 mV. Therefore, the ac current through the resistor is sinusoidal with a peak of

$$I_P = \frac{10 \text{ mV}}{10 \text{ k}\Omega} = 1 \ \mu\text{A}$$

The total current through the resistor is the algebraic sum of the dc and ac currents. Figure 5-1d illustrates this total current. It has an average or dc value of 1 mA. Superimposed on this average value is a sinusoidal component with a peak of 1 μA. On the positive peak, the total current is 1.001 mA. On the negative peak it is 0.999 mA.

Coupling capacitors

Transistor amplifiers often use capacitors to couple the ac signal but block the dc component. In other words, a large enough capacitor has a very low reactance to ac signals. Because of this, the capacitor will pass the ac component but stop the dc component. In drawing the dc and ac equivalent circuits, remember the following:

1. Direct current cannot exist in a capacitor; therefore, all capacitors look like *open circuits* in dc equivalent circuits.

75

Transistor Circuit Approximations

2. When ac signals are involved, capacitors usually couple the signal and are large enough to appear as *short circuits* in the ac equivalent circuit. Unless otherwise indicated, we automatically will assume all capacitors look like ac shorts.

Notation

To keep the dc and ac components distinct, we will use the following rules:

1. All dc and fixed quantities have capital letters. For instance, to represent a dc voltage and current, we will use V and I.
2. All ac and varying quantities have lowercase letters. For ac voltage and current, we will use v and i.

EXAMPLE 5-1

In Fig. 5-2a, what is the total voltage v_T produced by the dc and ac sources?

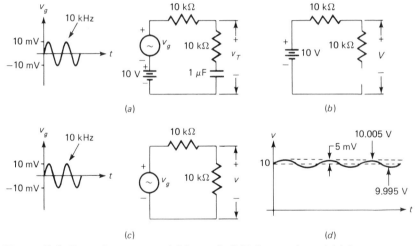

Figure 5-2 Equivalent circuits. *(a)* Original. *(b)* DC equivalent. *(c)* AC equivalent. *(d)* Total waveform.

SOLUTION

To get the dc equivalent circuit, reduce the ac source to zero and open the capacitor as shown in Fig. 5-2b. With no current in the circuit, the dc source voltage must appear across the output:

$$V = 10\text{ V}$$

To get the ac equivalent circuit, reduce the dc source to zero and short the capacitor (Fig. 5-2c). Because of the 2:1 voltage divider, the ac output v is a sine wave with a peak of 5 mV.

The total voltage v_T is the sum of dc and ac voltages. Figure 5-2d shows how it looks. The total voltage has a dc or average value of 10 V. Superimposed on this average value is a sinusoidal component with a peak of 5 mV. On the positive peak, the total voltage is 10.005 V and on the negative peak it is 9.995 V.

EXAMPLE 5-2

Find the total voltage v_T across the 10-Ω resistor of Fig. 5-3a.

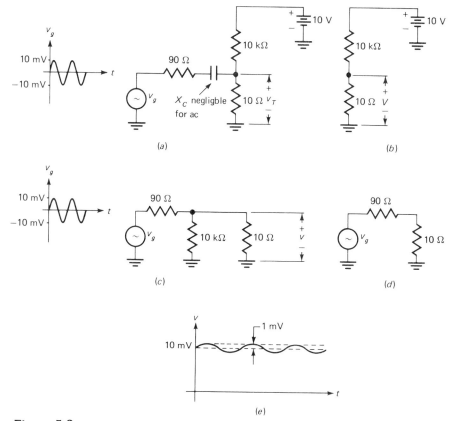

Figure 5-3

Transistor Circuit Approximations

SOLUTION

After reducing the ac source to zero and opening the capacitor, we get the dc equivalent circuit of Fig. 5-3b. With the voltage-divider theorem, the dc voltage across the 10 Ω is

$$V = \frac{10\ \Omega}{10{,}010\ \Omega} \times 10\ \text{V} = 9.99\ \text{mV} \cong 10\ \text{mV}$$

Next, reduce the dc source to zero and short the capacitor to get the ac equivalent circuit of Fig. 5-3c. 10 Ω in parallel with 10 kΩ is approximately 10 Ω. Therefore, the ac equivalent circuit simplifies to Fig. 5-3d. Because of the 10:1 voltage divider, the ac voltage v is a sine wave with a peak of 1 mV.

The total voltage produced by both sources acting simultaneously is the sum of the ac and dc components. Figure 5-3e shows this total voltage. It has an average value of 10 mV. Superimposed on this average value is a sine wave with a peak of 1 mV.

5-2 THE AC RESISTANCE OF A DIODE

To apply the superposition theorem to transistor amplifiers, we must discuss how a diode acts as far as small ac signals are concerned.

Small ac source

In Fig. 5-4a, a dc source is in series with a *small* ac source. Ideally, the diode appears shorted because it is forward-biased. The dc source sets up an average current of 1 mA. The ac source produces a sinusoidal variation with a peak of 0.1 μA. Therefore, the total current through the diode varies sinusoidally between 0.9999 and 1.0001 mA.

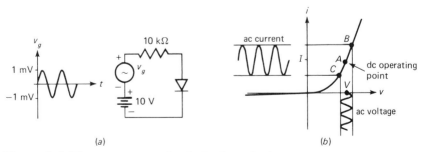

(a)   (b)

Figure 5-4 The ac resistance of a diode above the knee.

Figure 5-4*b* shows the diode voltage and current. The dc operating point is *A*. Because the ac changes are small, the arc between points *B* and *C* is approximately *linear*. This means ac voltage and current are directly proportional for small changes. For instance, if we reduce the ac voltage by a factor of 2 in Fig. 5-4*b*, the ac current is reduced by a factor of 2.

AC resistance

The linear relation between ac current and voltage means a diode looks like a resistance to small ac signals. In other words, diodes become resistors in the ac equivalent circuit, provided the ac signal is small. With calculus it is possible to derive this useful approximation for the ac resistance of a diode:

$$r_{ac} \cong \frac{25 \text{ mV}}{I} \tag{5-1}$$

where *I* is the *dc current* through the diode.

For instance, in Fig. 5-4*a* the dc source sets up a dc current of 1 mA through the diode. Therefore, the diode has an ac resistance of

$$r_{ac} \cong \frac{25 \text{ mV}}{1 \text{ mA}} = 25 \text{ }\Omega$$

This means the diode looks like a resistance of 25 Ω to the small ac signal driving it.

EXAMPLE 5-3

The dc current through a diode is 0.2 mA. What is the ac resistance of the diode?

SOLUTION

With Eq. (5-1),

$$r_{ac} \cong \frac{25 \text{ mV}}{I} = \frac{25 \text{ mV}}{0.2 \text{ mA}} = 125 \text{ }\Omega$$

EXAMPLE 5-4

If the dc current through a diode changes from 0.2 mA to 5 mA, what is the new value of ac resistance?

79

SOLUTION

Equation (5-1) gives

$$r_{ac} \cong \frac{25 \text{ mV}}{I} = \frac{25 \text{ mV}}{5 \text{ mA}} = 5 \text{ } \Omega$$

As you see, the dc current through a diode is important because it determines how much ac resistance a diode has.

5-3 APPLYING SUPERPOSITION TO DIODE CIRCUITS

As a preparation for transistor amplifiers, we need to analyze diode circuits with large dc sources and small ac sources. For such circuits, here is how we will apply the superposition theorem:

1. In the dc equivalent circuit, idealize the diode or use the second approximation to calculate the dc current through the diode. Also, work out any other dc currents or voltages needed to solve the problem.
2. When you draw the ac equivalent circuit, replace the diode by its ac resistance. Then, calculate any ac currents or voltages needed to solve the problem.
3. Add the dc and ac components to get the total current or voltage that you are interested in.

EXAMPLE 5-5

Find the total current i_T through the diode of Fig. 5-5a.

SOLUTION

Figure 5-5b shows the dc equivalent circuit. Ideally, the diode acts like a closed switch; therefore, the dc current through the diode is

$$I \cong \frac{10 \text{ V}}{10 \text{ k}\Omega} = 1 \text{ mA}$$

With Eq. (5-1), the diode's ac resistance is

$$r_{ac} \cong \frac{25 \text{ mV}}{I} = \frac{25 \text{ mV}}{1 \text{ mA}} = 25 \text{ } \Omega$$

Figure 5-5c is the ac equivalent circuit. We have replaced the diode by a resistance of 25 Ω. Notice how the shorted dc source makes the 10 kΩ

80

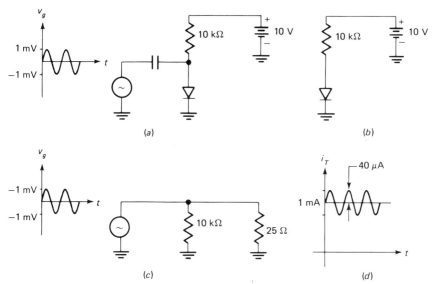

Figure 5-5

appear in parallel with the 25 Ω. The small ac source produces a sinusoidal current through the diode. This current has a peak value of

$$I_P = \frac{1 \text{ mV}}{25 \text{ } \Omega} = 40 \text{ } \mu\text{A}$$

The total current through the diode is the sum of the dc and ac currents. Figure 5-5d shows how this current looks. It has a dc value of 1 mA. Superimposed on this average value is a sine wave with a peak of 40 μA. Therefore, the total diode current is a sinusoidal variation between 0.96 and 1.04 mA.

GLOSSARY

ac equivalent circuit This is what you use to find ac currents and voltages. You get this equivalent circuit by reducing dc sources to zero, shorting coupling capacitors, and replacing diodes by their ac resistances.

ac resistance Whenever the ac current is much smaller than the dc current through a diode, the diode does not rectify the ac signal. Instead, it appears to be a resistor as far as the ac signal is concerned. This is what ac resistance means.

coupling capacitor This is a capacitor whose reactance is very small to

Transistor Circuit Approximations

the ac signal. Because of this, the coupling capacitor looks approximately like an ac short and a dc open. In other words, it transmits ac but blocks dc.

current source The current out of this source is independent of load resistance. Ideally, the Thevenin resistance of a current source is infinite and all the current from the current source must pass through the load resistor connected to it.

dc equivalent circuit This is what you use to find dc currents and voltages. You get this circuit by reducing all ac sources to zero and replacing all capacitors by open circuits. Either the ideal or second approximation is used for diodes in the dc equivalent circuit.

small signal To apply the superposition theorem to circuits containing dc and ac sources, the ac signal must be small. This means the ac current through the diode should be much smaller than the dc current. There is no hard-and-fast rule about the meaning of *much smaller*. As a guide, the peak ac current should be at least 10 times smaller than the dc current.

voltage source The voltage out of this source is independent of the load resistance. Ideally, the Thevenin resistance of a voltage source is zero and all the voltage from the voltage source appears across the load resistor connected to it.

SELF-TESTING REVIEW

Read each of the following and provide the missing words. Answers appear at the beginning of the next question.

1. In a network with two or more sources, the current or voltage for any component is the algebraic _____ of the effects produced by each source acting separately. To use one source at a time, reduce all other sources to _____. This means replacing voltage sources by _____ circuits and current sources by open circuits.

2. *(sum, zero, short)* In the dc equivalent circuit all capacitors appear like _____ circuits. When ac signals are involved, capacitors usually couple the signal and are large enough to appear as _____ circuits in the ac equivalent circuit.

3. *(open, short)* In the ac equivalent circuit a diode appears as a _____ when small ac signals are involved. This is because the diode current and voltage are directly proportional for small changes.

4. *(resistance)* The ac resistance of a diode equals 25 mV/I. If the dc current

82

through the diode is 1 mA, the ac resistance is _____ Ω. If the dc current is 5 mA, the ac resistance is _____ Ω.

5. *(25, 5)* When analyzing diode circuits driven by large dc and small ac sources, proceed as follows. In the dc equivalent circuit, idealize the diode or use the second approximation. In the ac equivalent circuit, replace the diode by its ac resistance.

PROBLEMS

5-1. How much dc current is there through the 50-Ω resistor of Fig. 5-6?

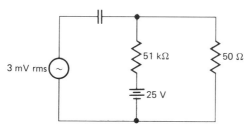

Figure 5-6

5-2. In Fig. 5-6, the source is sinusoidal with a value of 3 mV rms. What is the ac rms current through the 50 Ω?

5-3. Work out the total current through the 50-Ω resistor in Fig. 5-6.

5-4. Calculate the ac resistance of a diode for a dc current of 0.75 mA.

5-5. What is the ac resistance of a diode for each of the following dc currents:

a. 0.05 mA
b. 0.275 mA
c. 1.2 mA
d. 6.65 mA

5-6. In Fig. 5-7a, what is the dc current through the diode using the ideal approximation? The corresponding ac resistance of the diode?

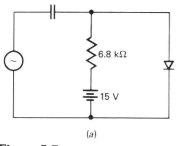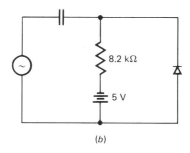

(a)   (b)

Figure 5-7

83

Transistor Circuit Approximations

5-7. Using the second approximation in Fig. 5-7b, calculate the dc current through the diode. Then, work out the ac resistance of the diode.

5-8. If the ac source of Fig. 5-7a has an rms value of 2 mV, what is the rms value of the ac current through the diode?

5-9. In Fig. 5-7b, the ac source has a peak-to-peak value of 5 mV. What is the rms value of the ac current through the diode?

5-10. In Fig. 5-7a, the ac source has a peak value of 1 mV. What is the total current through the diode?

5-11. The ac source of Fig. 5-7b has an rms value of 10 mV. Using the second approximation of the diode, what is the dc current through the diode? The total current through the diode?

Chapter 6

Common-Base
Approximations

This chapter introduces the transistor by discussing the *common-base* (CB) connection. Chapter 7 presents the *common-emitter* (CE) connection, and Chap. 8 examines the *common-collector* (CC) connection. Our first task is to find out how a transistor works. Then we can eliminate all minor effects to arrive at the *ideal transistor,* an approximation that simplifies analysis and design.

6-1 STRUCTURE

A *pnp* transistor has three doped regions as shown in Fig. 6-1*a*. The larger *p* region is called the *collector,* and the other *p* region is the *emitter.* In between the two *p* regions is an *n* region known as the *base.*

Figure 6-1*b* is an *npn* transistor. Again, there is an emitter, base, and collector. But this time, a *p* region is between two larger *n* regions.

Whether the transistor is *pnp* or *npn,* it resembles two back-to-back diodes. The one on the left is called the *emitter diode;* the one on the right is the

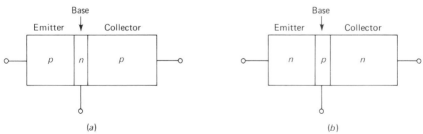

Figure 6-1 Transistors. *(a) pnp. (b) npn.*

collector diode. Since two types of charges are involved (holes and free electrons), *pnp* and *npn* transistors are classified as *bipolar* devices (*bipolar* means "two polarities").

Figure 6-2*a* is the schematic symbol for a *pnp* transistor. An arrowhead is on the emitter but not on the collector. This arrowhead, like the arrow in a diode, points in the easy direction of conventional flow. If you prefer using electron flow, the easy direction is the opposite way.

Similarly, Fig. 6-2*b* is the symbol for an *npn* transistor. This time, the arrowhead points out of the emitter, meaning the easy direction of *conventional flow* is *out of* the emitter. Therefore, the easy direction of *electron flow* is *into* the emitter.

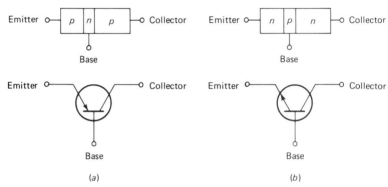

Figure 6-2 Schematic symbols. *(a) pnp. (b) npn.*

6-2 BIASING THE BIPOLAR TRANSISTOR

Figure 6-3*a* shows an *npn* transistor with a grounded base. Because the base is common to both the emitter and collector loops, this connection is known as the *common-base* connection. Sometimes, it is referred to as the grounded-base connection.

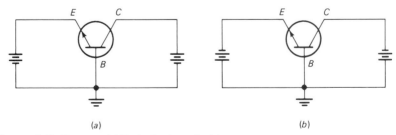

Figure 6-3 Biasing. *(a)* Both diodes off. *(b)* Both diodes on.

Both diodes reverse-biased

The dc sources of Fig. 6-3a reverse-bias the emitter and collector diodes. Aside from minority carriers and surface leakage, the current in each diode is small enough to ignore. In other words, with both diodes reverse-biased, nothing unusual happens.

Both diodes forward-biased

In Fig. 6-3b, the dc sources have been turned around to forward-bias both diodes. Now, we get a large current in each diode. This is exactly what we expect, so that nothing unusual happens when both diodes are forward-biased.

Emitter forward-biased and collector reverse-biased

Something new and different happens when we forward-bias the emitter diode and reverse-bias the collector diode as shown in Fig. 6-4. Our initial reaction says there should be a large emitter current and a small collector current. This is not what happens! Instead, there is a large emitter current and an almost equally large collector current.

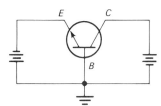

Figure 6-4 Emitter forward-biased and collector reverse-biased.

Why is the collector current large even though the collector diode is reverse-biased? In Fig. 6-5a, the n-type emitter has many free electrons. Because the emitter diode is forward-biased, these majority carriers are forced into the base region. Since the base is very thin and lightly doped, most of these free electrons have enough lifetime to diffuse through the base into the collector. The positive collector source voltage then attracts these free electrons, as shown in Fig. 6-5b.

The path of one electron

Let us follow one free electron all the way through the circuit of Fig. 6-5b. To begin with, a free electron leaves the negative terminal of the emitter

Transistor Circuit Approximations

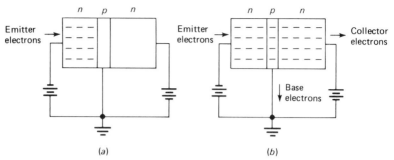

Figure 6-5 Emitter electrons captured by collector.

power supply. This free electron enters the emitter and flows toward the base. Once inside the base, the free electron becomes a minority carrier because it is now in *p*-type material. But here are two crucial points:

1. The base is very lightly doped.
2. The base is very thin.

Because of the light doping, the free electron has a very long lifetime; and since the base is so thin, the free electron has enough time to diffuse through the base into the collector.

Once the free electron is inside the collector, it again becomes a majority carrier because it is in *n*-type material. As a majority carrier, it flows through the collector into the positive terminal of the collector supply.

Main ideas

Here are the things to remember :

1. For normal operation of a transistor, forward-bias the emitter diode and reverse-bias the collector diode.
2. Collector current is almost equal to emitter current.
3. Base current is very small, equal to the difference of emitter and collector current.

6-3 COMMON-BASE COLLECTOR CURVES

Figure 6-6*a* shows adjustable dc sources driving an *npn* transistor. The voltage across the emitter diode is V_{BE}, and the voltage across the collector diode is V_{CB}. If we adjust the emitter supply to get 1 mA of emitter current, we

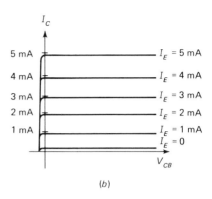

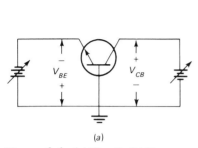

(a)   (b)

Figure 6-6 *(a)* Circuit. *(b)* Curves.

will find the collector current is approximately 1 mA. Even though we vary the collector supply, the collector current remains at approximately 1 mA. This ties in with the earlier explanation of transistor action. Almost all of the free electrons passing from emitter to base diffuse into the collector because the base is thin and lightly doped. The collector captures almost all of the free electrons whether V_{CB} is small or large.

Curves

Figure 6-6*b* summarizes the idea. When the emitter current is 1 mA, the collector current is approximately 1 mA for all collector voltages greater than zero. If we adjust the emitter supply to get 2 mA of emitter current, the collector current automatically increases to approximately 2 mA. Once the emitter current has been set by the emitter supply, changing the collector supply has almost no effect.

Controlled current source

The curves of Fig. 6-6*b* imply that the collector part of a transistor acts like a controlled *current source*. As discussed in basic circuit books, a current source is a device with infinite Thevenin resistance. Because of this, it puts out a current that does not depend on the voltage across its terminals. In Fig. 6-6*b*, when $I_E = 1$ mA, $I_C \cong 1$ mA no matter what the value of V_{CB}.

The current source is controlled by the emitter current. When we change the emitter current, the collector current changes to approximately the same value. Again, this collector current is not influenced by collector voltage. No matter what the value of V_{CB}, the collector current remains approximately constant for a fixed emitter current.

89

Cutoff current

In Fig. 6-6b, a small collector current exists even when $I_E = 0$. It consists of minority carriers and surface leakage. Although exaggerated in Fig. 6-6b, it is actually very small and we can ignore it in most applications.

6-4 ALPHA AND BETA

Collector current almost equals emitter current. To pin down how close collector current is to emitter current, we will define the *dc alpha* of a transistor. Furthermore, collector current is much greater than base current. The *dc beta* is a measure of how much greater. Both the dc alpha and beta will be used in our later work.

Alpha

The dc alpha of a transistor is defined as the dc collector current divided by the dc emitter current. In symbols,

$$\alpha_{dc} = \frac{I_C}{I_E} \qquad\qquad \textbf{(6-1)}$$

For instance, if the dc collector current is 2.94 mA when the dc emitter current is 3 mA,

$$\alpha_{dc} = \frac{2.94 \text{ mA}}{3 \text{ mA}} = 0.98$$

The α_{dc} is a measure of the quality of a transistor. The higher the α_{dc}, the better the transistor action because collector current more closely equals emitter current. Ideally, $\alpha_{dc} = 1$, because then all carriers from the emitter pass through the base to the collector. Almost all transistors have an α_{dc} between 0.95 and 1. This means dc collector current is between 95 and 100 percent of dc emitter current.

Beta

The dc beta of a transistor is defined as the ratio of dc collector current to dc base current. In symbols,

$$\beta_{dc} = \frac{I_C}{I_B} \qquad\qquad \textbf{(6-2)}$$

90

For example, if the dc collector current is 5 mA when the dc base current is 0.025 mA,

$$\beta_{dc} = \frac{5 \text{ mA}}{0.025 \text{ mA}} = 200$$

The dc beta is an indication of how well the transistor works. The higher β_{dc} is, the better the transistor action. In almost any transistor, less than 5 percent of the emitter electrons recombine with base holes to produce I_B; therefore, β_{dc} is almost always greater than 20. Usually, it is from 50 to 300. And some transistors have β_{dc} as high as 1000.

In another system of analysis called *h parameters*, β_{dc} is known as the dc current gain and is designated h_{FE}. In other words, $\beta_{dc} = h_{FE}$. This is important to remember because data sheets for transistors usually list the value of h_{FE}.

Relationship between α_{dc} and β_{dc}

Data sheets seldom list the value of α_{dc}. In case you need it, here is the relationship between α_{dc} and β_{dc}:

$$\alpha_{dc} = \frac{\beta_{dc}}{\beta_{dc} + 1} \qquad \textbf{(6-3)}$$

For instance, if a data sheet shows $h_{FE} = 250$, then $\beta_{dc} = 250$ and

$$\alpha_{dc} = \frac{250}{250 + 1} = 0.996$$

6-5 THE IDEAL TRANSISTOR

In a perfect transistor, α_{dc} is equal to unity. This means all emitter carriers entering the base region go on to the collector. Therefore, the collector current of an ideal transistor equals the emitter current.

Figure 6-7a shows the equivalent circuit for an ideal *npn* transistor. The emitter diode acts like any forward-biased diode. Because of transistor action, however, the collector diode acts like a current source. The current-source arrow points in the easy direction for conventional current; therefore, electrons flow easily in the opposite direction. This current source is controlled by the emitter current.

91

Transistor Circuit Approximations

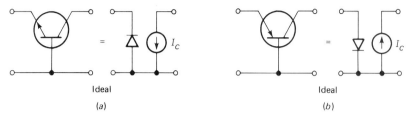

Figure **6-7** Ideal equivalent circuits. *(a) npn (b) pnp.*

Figure 6-7*b* is the equivalent circuit for an ideal *pnp* transistor. Again, the emitter diode acts like an ordinary diode, but the collector diode acts like a controlled current source.

EXAMPLE 6-1

What is the ideal dc collector current in Fig. 6-8*a?* The value for a V_{BE} drop of 0.7 V?

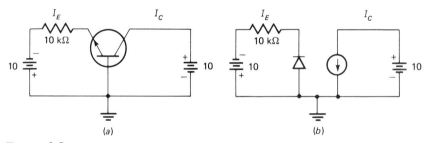

Figure 6-8

SOLUTION

Figure 6-8*b* shows the equivalent circuit. If we ignore the V_{BE} drop across the emitter diode, the dc emitter current is

$$I_E = \frac{10\text{ V}}{10\text{ k}\Omega} = 1\text{ mA}$$

The dc collector current ideally equals the dc emitter current; therefore,

$$I_C \cong 1\text{ mA}$$

To a second approximation, we can take the drop across the emitter diode into account and get a dc emitter current of

92

$$I_E = \frac{10\text{ V} - 0.7\text{ V}}{10\text{ k}\Omega} = 0.93\text{ mA}$$

The dc collector current is approximately equal to I_E:

$$I_C \cong 0.93\text{ mA}$$

EXAMPLE 6-2

Allow 0.7 V for the V_{BE} drop in Fig. 6-9a and calculate the dc collector-base voltage V_{CB}.

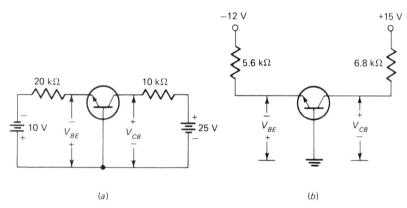

(a)   (b)

Figure 6-9

SOLUTION

The dc emitter current is

$$I_E = \frac{10\text{ V} - 0.7\text{ V}}{20\text{ k}\Omega} = 0.465\text{ mA}$$

The dc collector current has approximately the same value:

$$I_C \cong 0.465\text{ mA}$$

The dc voltage between the collector and the base equals the collector supply voltage minus the drop across the collector resistor:

$$V_{CB} = 25 - 10,000\ I_C$$
$$= 25 - 10,000(0.465 \times 10^{-3})$$
$$= 20.35\text{ V}$$

93

Transistor Circuit Approximations

EXAMPLE 6-3

Figure 6-9*b* is a simplified way to draw a transistor circuit. A negative supply forward-biases the emitter diode, and a positive supply reverse-biases the collector diode. Allowing 0.7 V for V_{BE}, what does V_{CB} equal?

SOLUTION

$$I_E = \frac{12\ \text{V} - 0.7\ \text{V}}{5.6\ \text{k}\Omega} = 2.02\ \text{mA}$$

The collector-base voltage is

$$\begin{aligned} V_{CB} &= 15 - 6800 I_C \\ &= 15 - 6800(2.02 \times 10^{-3}) \\ &= 1.26\ \text{V} \end{aligned}$$

6-6 ANALYZING COMMON-BASE AMPLIFIERS

Figure 6-10*a* is a common-base amplifier. The V_{EE} source forward-biases the emitter diode, and the V_{CC} source reverse-biases the collector diode. Typically, the ac source is much smaller than the dc sources. Therefore, it produces small fluctuations in transistor currents and voltages. As will be

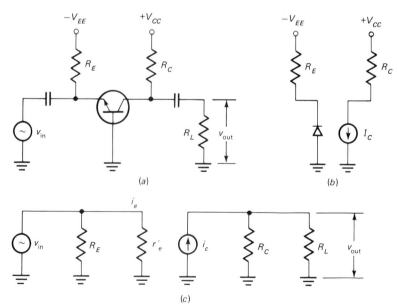

Figure 6-10 *(a)* Original circuit. *(b)* DC equivalent. *(c)* AC equivalent.

94

shown, the ac output voltage is amplified because it is larger than the ac input voltage.

DC equivalent circuit

You get the dc equivalent circuit by reducing all ac sources to zero and opening all capacitors. In the dc equivalent circuit of Fig. 6-10*b*, we can calculate the dc emitter current by using either the ideal or the second approximation for the emitter diode. The dc collector current is approximately equal to this dc emitter current. And the dc collector-base voltage equals the collector supply voltage minus the drop across the collector resistor. As a formula,

$$V_{CB} = V_{CC} - I_C R_C \qquad \textbf{(6-4)}$$

AC equivalent circuit

To get the ac equivalent circuit, reduce all dc sources to zero, short all coupling capacitors, and replace the emitter diode by its ac resistance. In transistor-circuit analysis, r_e' represents the ac resistance of the emitter diode. To calculate r_e', use

$$r_e' = \frac{25\ \text{mV}}{I_E} \qquad \textbf{(6-5)}$$

where I_E is the dc emitter current.

Figure 6-10*c* is the ac equivalent circuit. Notice how the emitter diode is replaced by its ac resistance. Also notice how the collector diode is replaced by an ac current source. To a close approximation, the small changes in collector current equal the small changes in emitter current. In other words, ac collector current approximately equals ac emitter current:

$$i_c \cong i_e$$

During the positive half cycle of input voltage, the conventional direction for ac collector current is upward as shown by the current-source arrow.

Voltage gain

Since the ac input voltage is across the emitter diode, the ac emitter current is

$$i_e = \frac{v_{in}}{r_e'}$$

which means $$v_{in} = i_e r_e'$$ **(6-6)**

The ac collector current drives the parallel connection of R_C and R_L; therefore,

$$v_{out} = i_c(R_C \parallel R_L)$$ **(6-7)**

where $R_C \parallel R_L$ is the equivalent parallel resistance of R_C and R_L.

Voltage gain A is defined as the ratio of ac output voltage to ac input voltage. With Eqs. (6-6) and (6-7), the voltage gain of a common-base amplifier is

$$A = \frac{v_{out}}{v_{in}} = \frac{i_c(R_C \parallel R_L)}{i_e r_e'}$$

Since $i_c \cong i_e$, this reduces to

$$A = \frac{r_C}{r_e'}$$ **(6-8)**

where $r_C = R_C \parallel R_L$. Incidentally, voltage gain is sometimes symbolized by A_v. For this reason, you may see the foregoing equation expressed as $A_v = r_C/r_e'$.

EXAMPLE 6-4

Calculate the ac emitter current and the ac output voltage in Fig. 6-11.

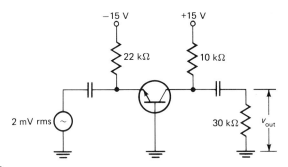

Figure 6-11

SOLUTION

First, get the dc emitter current. Allowing 0.7 V for the drop across the emitter diode,

96

$$I_E = \frac{15\,V - 0.7\,V}{22\,k\Omega} = 0.65\,mA$$

Next, get the ac emitter resistance:

$$r_e' = \frac{25\,mV}{0.65\,mA} = 38.5\,\Omega$$

Now, the ac emitter current equals the ac input voltage divided by the ac emitter resistance:

$$i_e = \frac{v_{in}}{r_e'} = \frac{2\,mV}{38.5\,\Omega} = 0.0519\,mA$$

The ac collector current approximately equals 0.0519 mA, which gives an ac output voltage of

$$
\begin{aligned}
v_{out} &= 0.0519\,mA \times (10\,k\Omega \parallel 30\,k\Omega) \\
&= 0.0519\,mA \times 7.5\,k\Omega \\
&= 0.389\,V
\end{aligned}
$$

EXAMPLE 6-5

Calculate the voltage gain of Fig. 6-11.

SOLUTION

In the preceding example, we found an ac output voltage of 0.389 V for an ac input voltage of 2 mV. Voltage gain is defined as the ratio of v_{out} to v_{in}; therefore,

$$A = \frac{v_{out}}{v_{in}} = \frac{0.389\,V}{2\,mV} = 195$$

This means the ac output voltage is 195 times larger than the ac input voltage. An alternative way to find the voltage gain is with Eq. (6-8):

$$A = \frac{r_C}{r_e'} = \frac{10\,k\Omega \parallel 30\,k\Omega}{38.5\,\Omega} = \frac{7.5\,k\Omega}{38.5\,\Omega} = 195$$

EXAMPLE 6-6

What is the voltage gain in Fig. 6-12a? If $v_{in} = 1.5$ mV, what does v_{out} equal?

97

Transistor Circuit Approximations

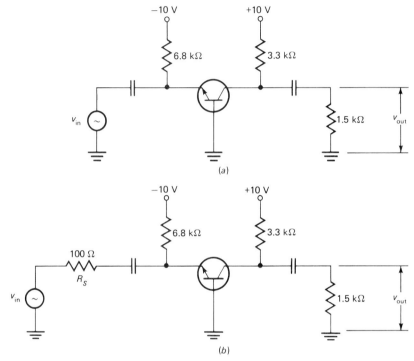

Figure 6-12

SOLUTION

Allowing 0.7 V across the emitter diode, the dc emitter current is

$$I_E = \frac{10\,V - 0.7\,V}{6.8\,k\Omega} = 1.37\,mA$$

and the ac emitter resistance is

$$r_e' = \frac{25\,mV}{1.37\,mA} = 18.2\,\Omega$$

The voltage gain is

$$A = \frac{r_C}{r_e'} = \frac{3.3\,k\Omega \parallel 1.5\,k\Omega}{18.2\,\Omega} = \frac{1.03\,k\Omega}{18.2\,\Omega} = 56.6$$

With a v_{in} of 1.5 mV, the ac output voltage is

$$v_{out} = Av_{in} = 56.6 \times 1.5\,mV = 84.9\,mV$$

98

EXAMPLE 6-7

Figure 6-12*b* is similar to Fig. 6-12*a*, except for the source resistance R_S. What is the voltage gain?

SOLUTION

Source resistance R_S is in series with r_e' in the ac equivalent circuit. If you review the derivation of Eq. (6-8), you can see that R_S has to be added to r_e'. In other words, the voltage gain with a source resistance is

$$A = \frac{r_C}{R_S + r_e'} \tag{6-9}$$

As mentioned earlier, A_v is an alternative symbol for voltage gain. Equation (6-9) therefore may be written as $A_v = r_C/(R_S + r_e')$.
   With the circuit values of Fig. 6-12*b*,

$$A = \frac{3.3 \text{ k}\Omega \parallel 1.5 \text{ k}\Omega}{100 \text{ }\Omega + 18.2 \text{ }\Omega} = 8.71$$

This demonstrates the effect of source resistance; it lowers the overall voltage gain of the circuit. For a v_{in} of 1.5 mV, the ac output voltage is

$$v_{out} = A v_{in} = 8.71 \times 1.5 \text{ mV} = 13.1 \text{ mV}$$

(Compare this to the output voltage in Example 6-6.)

GLOSSARY

alpha The ratio of collector current to emitter current.

base The middle region of a transistor. It is very thin and lightly doped.

beta The ratio of collector current to base current.

collector The largest of the transistor regions because it has to dissipate the most power.

common base One of the ways to connect a transistor. The base is common to both the emitter and the collector loops. Common-base is synonymous with grounded-base.

coupling capacitor A capacitor large enough to appear as a very low reactance to the ac signal. Because of this, a coupling capacitor transmits the ac component but blocks the dc component.

99

Transistor Circuit Approximations

emitter The most heavily doped of the three transistor regions. It supplies the majority carriers that move through the base to the collector.

emitter resistance The ac resistance of the emitter diode. This important quantity is designated r'_e. As a theoretical guide, it's equal to 25 mV divided by I_E. In practice, r'_e may range from 25 mV/I_E to 50 mV/I_E.

lifetime The average amount of time a minority carrier exists before recombining.

voltage gain The ratio of the ac output voltage to the ac input voltage.

SELF-TESTING REVIEW

Read each of the following and provide the missing words. Answers appear at the beginning of the next question.

1. The three regions of a transistor are the emitter, base, and collector. The base is very _____ and lightly _____.

2. *(thin, doped)* The correct way to operate a transistor is with the emitter diode forward-biased and the _____ diode reverse-biased. In this case, the collector current is almost _____ to the emitter current. The base current is very _____.

3. *(collector, equal, small)* Ideally, the emitter diode acts like a forward-biased _____, and the collector diode acts like a _____ source. Changing the collector voltage has little effect on collector current. To change collector current we have to change the emitter current.

4. *(diode, current)* The dc alpha equals the ratio of dc collector current to dc emitter current. The dc beta is the ratio of the dc collector current to dc _____ current. Typically, the dc alpha is approximately _____, and the dc beta is from 50 to 300.

5. *(base, unity)* Analysis of a transistor amplifier involves the dc equivalent circuit and the ac equivalent circuit. In the ac equivalent circuit, the emitter diode acts like a _____ and is approximately equal to 25 mV/I_E.

6. *(resistance)* Voltage gain is defined as the ratio of ac output voltage to ac input voltage. If you know the value of A, you can find v_{out} by using $v_{out} = Av_{in}$.

PROBLEMS

6-1. When the dc collector current is 4.9 mA, the dc emitter current is 5 mA. What does α_{dc} equal? What does β_{dc} equal?

100

6-2. A transistor data sheet lists an h_{FE} of 175. What are the values of α_{dc} and β_{dc}?

6-3. A 2N3904 is a silicon *npn* transistor with a minimum h_{FE} of 100 and a maximum h_{FE} of 300. What are the minimum and maximum values of α_{dc} for this transistor?

6-4. What is the dc emitter current in Fig. 6-13a? The dc collector-base voltage?

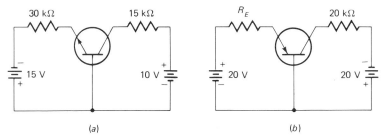

(a)   (b)

Figure 6-13

6-5. If $R_E = 39$ kΩ in Fig. 6-13b, what does the dc emitter current equal? The dc collector-base voltage?

6-6. In Fig. 6-13b, $R_E = 22$ kΩ. Find the values of the dc emitter current and the dc collector-base voltage.

6-7. What is the dc emitter current in Fig. 6-14? The ac resistance of the emitter diode?

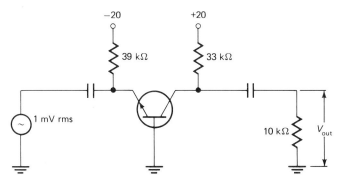

Figure 6-14

6-8. In Fig. 6-14, what does the voltage gain equal? The ac output voltage?

6-9. What does the dc collector-base voltage equal in Fig. 6-14? The ac collector-base voltage? The total collector-base voltage?

101

Chapter 7

Common-Emitter Approximations

The *common-emitter* (CE) connection is the most widely used. This chapter discusses different ways to bias a CE connection and then tells you how to analyze CE amplifiers.

7-1 THE CE CONNECTION

A bipolar transistor can be connected with the emitter grounded instead of the base, as shown in Fig. 7-1a. In this circuit, the emitter is common to the base and collector loops. This is why it is called the common-emitter connection. Sometimes, it is referred to as the grounded-emitter connection.

Collector curves

During this discussion, assume the transistor of Fig. 7-1a has a β_{dc} of 100. If the base current is adjusted to 0.01 mA, the collector current becomes 1 mA. Varying the collector supply voltage has little effect on the collector current because almost all free electrons from the emitter pass through the base region to the collector.

If the base current is changed to 0.02 mA, the collector current automatically changes to 2 mA. Again, the collector current is approximately 100 times greater than the base current. And again, varying the collector supply voltage has almost no effect on the collector current.

Figure 7-1b summarizes the transistor action for different base currents. For a given base current, an increase in collector voltage produces only a slight increase in collector current. To a first approximation, the collector current is approximately constant and equal to 100 times the base current.

102

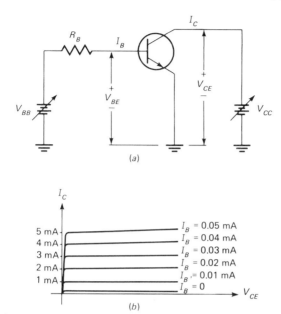

Figure 7-1 *(a)* Circuit. *(b)* Transistor curves.

Saturation, breakdown, and active regions

Collector current is constant over a large range of V_{CE}. But, as shown in Fig. 7-1b, when V_{CE} approaches zero, collector current drops off. The almost vertical region where the collector current decreases rapidly is called the *saturation region*. In this region, V_{CE} is typically only a few tenths of a volt. In the saturation region, the collector diode is coming out of reverse bias and is no longer collecting all the free electrons that it can. To avoid the saturation region, V_{CE} should be greater than 1 V (valid for most transistors).

On the other hand, if V_{CE} is too large, the collector diode will break down (not shown in Fig. 7-1b). For instance, a 2N3904 has a collector breakdown voltage of 40 V. For normal operation, V_{CE} should be less than 40 V.

The *active region* is where the collector curves are almost horizontal. In the active region, the collector current equals approximately 100 times the base current (Fig. 7-1b). In other words, the active region is all the area between the saturation and breakdown regions. With a 2N3904, this means the active region occurs for V_{CE} between 1 V and 40 V.

The ideal CE approximation

Figure 7-2 is the ideal approximation for an *npn* transistor with a CE connection. The base-emitter part of the transistor acts like a forward-biased diode

103

Transistor Circuit Approximations

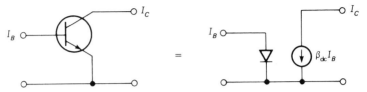

Figure 7-2 Ideal dc equivalent circuit.

with a current of I_B. The collector-emitter part of the transistor acts like a current source with a value of $\beta_{dc} I_B$. With a CE connection, the collector current source is controlled by base current. When we change the base current, the collector current automatically changes to a new value equal to $\beta_{dc} I_B$.

7-2 BASE BIAS

Figure 7-3a shows *base bias,* the simplest way to bias a CE connection.

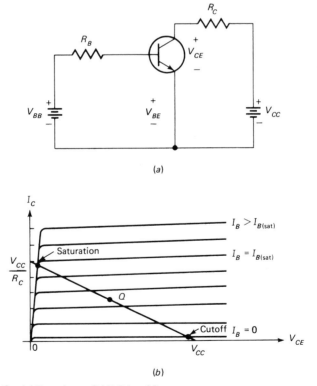

(a)

(b)

Figure 7-3 *(a)* Base bias. *(b)* DC load line.

104

A voltage source V_{BB} forward-biases the emitter diode through a current-limiting resistor R_B. The dc base current through this resistor is

$$I_B = \frac{V_{BB} - V_{BE}}{R_B} \qquad \text{(7-1)}$$

As usual, $V_{BE} = 0.7$ V for silicon transistors (0.3 V for germanium).

DC load line

Voltage source V_{CC} reverse-biases the collector diode through resistor R_C. The collector-emitter voltage V_{CE} equals the collector supply voltage minus the drop across the collector resistor. In symbols,

$$V_{CE} = V_{CC} - I_C R_C \qquad \text{(7-2)}$$

For a particular circuit, V_{CC} and R_C are constants; V_{CE} and I_C are variables.

Equation (7-2) is a linear equation. When graphed, it is a straight line superimposed on the collector curves (see 7-3b). The line intersects the horizontal axis at V_{CC} and the vertical axis at V_{CC}/R_C. This line is called the *dc load line* because it contains all possible dc operating points for the transistor.

Cutoff and saturation

The point where the dc load line intersects the $I_B = 0$ curve is known as *cutoff*. At this point, base current is zero and collector current is extremely small. At cutoff, the emitter diode comes out of forward bias and normal transistor action is lost.

Near the upper end of the load line is *saturation*. At this point, the collector diode comes out of reverse bias and collector current reaches a *maximum* value. Because V_{CE} is almost zero in the saturation region, the collector current is approximately

$$I_{C(\text{sat})} \cong \frac{V_{CC}}{R_C} \qquad \text{(7-3)}$$

The minimum base current that produces saturation is

$$I_{B(\text{sat})} = \frac{I_{C(\text{sat})}}{\beta_{\text{dc}}} \qquad \text{(7-4)}$$

105

Transistor Circuit Approximations

If the base current is greater than $I_{B(\text{sat})}$, the collector current *cannot* increase because the collector diode is no longer reverse-biased.

The Q point

All points on the load line between cutoff and saturation are the active region of the transistor. In the active region the emitter diode is forward-biased and the collector diode is reverse-biased. When used in an amplifier, the transistor should be biased somewhere near the middle of the load line at point Q. This point is called the *quiescent* (at-rest) point because it is the operating point when no ac signal is present. When an ac signal is present, it causes fluctuations above and below the Q point.

EXAMPLE 7-1

The 2N3904 of Fig. 7-4a has a β_{dc} of 125. What is the dc voltage between the collector and emitter?

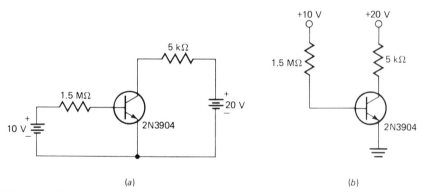

(a)   (b)

Figure 7-4

SOLUTION

Start by getting the dc base current. It equals the voltage across the base resistor divided by its resistance:

$$I_B = \frac{V_{BB} - V_{BE}}{R_B} = \frac{10 \text{ V} - 0.7 \text{ V}}{1.5 \text{ M}\Omega} = 6.2 \text{ } \mu\text{A}$$

The dc collector current is

$$I_C = \beta_{\text{dc}} I_B = 125 \times 6.2 \text{ } \mu\text{A} = 0.775 \text{ mA}$$

106

The dc collector-emitter voltage equals the collector supply voltage minus the drop across the collector resistor:

$$V_{CE} = V_{CC} - I_C R_C = 20\ \text{V} - (0.775\ \text{mA} \times 5\ \text{k}\Omega) = 16.1\ \text{V}$$

Figure 7-4b is the same circuit drawn in a simpler form. In this circuit, the negative terminals of the dc sources are grounded to get a complete path for current. From now on, we will use this simplified form.

EXAMPLE 7-2

The 2N4401 of Fig. 7-5 has a β_{dc} of 80. Draw the dc load line.

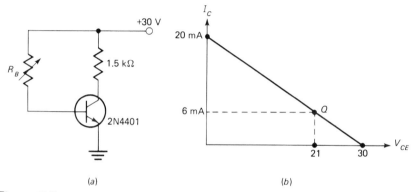

(a)   (b)

Figure 7-5

SOLUTION

By decreasing R_B, we eventually get enough base current to saturate the transistor. The collector current at this point is

$$I_{C(\text{sat})} \cong \frac{V_{CC}}{R_C} = \frac{30\ \text{V}}{1.5\ \text{k}\Omega} = 20\ \text{mA}$$

If we open R_B, base current drops to zero and the transistor goes into cutoff. In this case, all the supply voltage appears across the collector-emitter terminals and

$$V_{CE} = 30\ \text{V}$$

Figure 7-5b shows the dc load line. As you see, saturation current is 20 mA and cutoff voltage is 30 V.

107

Transistor Circuit Approximations

EXAMPLE 7-3

What are the coordinates of the Q point in Fig. 7-5a when $R_B = 390$ kΩ? (Use $\beta_{dc} = 80$.)

SOLUTION

The 2N4401 is a silicon transistor. The dc base current is

$$I_B = \frac{V_{BB} - V_{BE}}{R_B} = \frac{30 \text{ V} - 0.7 \text{ V}}{390 \text{ k}\Omega} = 75.1 \text{ } \mu A$$

The dc collector current is

$$I_C = \beta_{dc} I_B = 80 \times 75.1 \text{ } \mu A = 6 \text{ mA}$$

The dc collector-emitter voltage is

$$V_{CE} = V_{CC} - I_C R_C = 30 \text{ V} - (6 \text{ mA} \times 1.5 \text{ k}\Omega) = 21 \text{ V}$$

Figure 7-5b shows the Q point. It has coordinates of 6 mA and 21 V. The Q point lies on the dc load line, because the load line contains all possible operating points. If we were to change the value of R_B, the Q point would shift to another point on the dc load line.

EXAMPLE 7-4

The TIL222 of Fig. 7-6 is a green LED with a forward voltage of 2.3 V when conducting. If the 2N3904 has a β_{dc} of 150, what is the LED current when the transistor is saturated? The value of V_{IN} that just saturates the transistor?

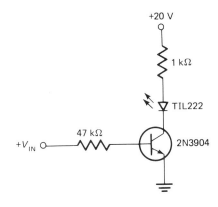

Figure 7-6

108

SOLUTION

When the transistor is saturated, its collector is approximately grounded because V_{CE} is almost zero. The voltage across the 1 kΩ equals the supply
voltage minus the drop across the LED; therefore,

$$I_{LED} \cong \frac{V_{CC} - V_{LED}}{R_C} = \frac{20 \text{ V} - 2.3 \text{ V}}{1 \text{ k}\Omega} = 17.7 \text{ mA}$$

With this current, the LED emits a bright green light.
The dc base current that just saturates the transistor is

$$I_{B(\text{sat})} = \frac{I_{C(\text{sat})}}{\beta_{dc}} = \frac{17.7 \text{ mA}}{150} = 0.118 \text{ mA}$$

The corresponding input voltage is

$$V_{\text{IN}} = I_B R_B + V_{BE} = (0.118 \text{ mA} \times 47 \text{ k}\Omega) + 0.7 = 6.25 \text{ V}$$

As long as the input voltage is greater than or equal to 6.25 V, the dc base
current is large enough to saturate the transistor.

7-3 VOLTAGE-DIVIDER BIAS

Linear transistor circuits are those that always operate in the active region.
In other words, the transistors in linear circuits are never driven into saturation
or cutoff. Typically, the dc source sets up a Q point near the middle of
the load line, and a small ac source produces fluctuations in collector current
and voltage.

Figure 7-7a shows *voltage-divider bias,* the most widely used bias for linear
transistor circuits. Its name comes from the voltage divider formed by R_1
and R_2. The voltage across R_2 forward-biases the emitter diode and produces
a collector current that is almost unaffected by the value of β_{dc}. This is
the main reason for the great popularity of voltage-divider bias.

Emitter current

Voltage-divider bias works like this. The base current in Fig. 7-7a is small
compared to the current through R_1 and R_2. Because of this, the voltage
across R_2 is

$$V_2 \cong \frac{R_2}{R_1 + R_2} V_{CC} \tag{7-5}$$

109

Transistor Circuit Approximations

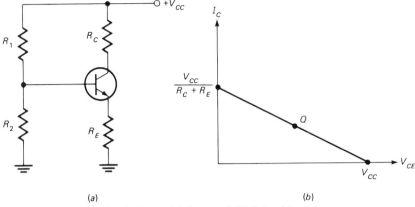

(a)   (b)

Figure 7-7 Voltage-divider bias. *(a)* Circuit. *(b)* DC load line.

The voltage across the emitter resistor equals V_2 minus the V_{BE} drop:

$$V_E = V_2 - V_{BE}$$

Therefore, the dc emitter current is

$$I_E = \frac{V_E}{R_E} = \frac{V_2 - V_{BE}}{R_E} \qquad \textbf{(7-6)}$$

As usual, collector current approximately equals emitter current, so once you have calculated the emitter current, you have the collector current.

DC load line

When a transistor is saturated, V_{CE} is approximately zero, which is equivalent to saying the collector and emitter are shorted. In Fig. 7-7a, visualize the collector and emitter shorted. Then you can see that all the supply voltage is across the series connection of R_C and R_E. Because of this, the saturation current is

$$I_{C(\text{sat})} = \frac{V_{CC}}{R_C + R_E} \qquad \textbf{(7-7)}$$

On the other hand, when a transistor is cut off, its collector-emitter terminals appear open. In this case, all the supply voltage of Fig. 7-7a is across the collector-emitter terminals and

$$V_{CE} \cong V_{CC} \qquad \text{(cutoff)} \qquad \textbf{(7-8)}$$

110

Figure 7-7*b* shows the dc load line for voltage-divider bias. As usual, the *Q* point is near the middle of the load line to allow for ac fluctuations when an ac source is used. In this way, we can get small changes above and below the *Q* point without driving the transistor into saturation or cutoff.

Voltages

The dc collector-emitter voltage equals the supply voltage minus the drop across the collector and emitter resistors. Because collector and emitter current are approximately equal,

$$V_{CE} = V_{CC} - I_C(R_C + R_E) \qquad (7\text{-}9)$$

When troubleshooting, it's more convenient to measure voltages with respect to ground. The dc collector-ground voltage equals the supply voltage minus the drop across the collector resistor:

$$V_C = V_{CC} - I_C R_C \qquad (7\text{-}10)$$

As mentioned earlier, the dc emitter-ground voltage equals the voltage across R_2 minus the V_{BE} drop:

$$V_E = V_2 - V_{BE} \qquad (7\text{-}11)$$

EXAMPLE 7-5

In Fig. 7-8*a*, what is the dc voltage from collector to ground?

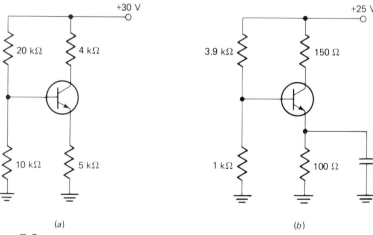

(a)   (b)

Figure 7-8

111

Transistor Circuit Approximations

SOLUTION

The voltage across the 10-kΩ base resistor is 10 V. The emitter diode drops approximately 0.7 V; therefore,

$$V_E = 10 \text{ V} - 0.7 \text{ V} = 9.3 \text{ V}$$

The emitter current is

$$I_E = \frac{V_E}{R_E} = \frac{9.3 \text{ V}}{5 \text{ k}\Omega} = 1.86 \text{ mA}$$

Since collector current approximately equals emitter current,

$$V_C = V_{CC} - I_C R_C = 30 \text{ V} - (1.86 \text{ mA} \times 4 \text{ k}\Omega) = 22.6 \text{ V}$$

EXAMPLE 7-6

Work out the dc collector-ground voltage in Fig. 7-8b.

SOLUTION

The capacitor is open to dc; therefore, it has nothing to do with the problem. The voltage across the 1-kΩ resistor is

$$V_2 = \frac{1 \text{ k}\Omega}{4.9 \text{ k}\Omega} 25 \text{ V} = 5.1 \text{ V}$$

The dc emitter-ground voltage is

$$V_E = 5.1 \text{ V} - 0.7 \text{ V} = 4.4 \text{ V}$$

The dc emitter current is

$$I_E = \frac{4.4 \text{ V}}{100 \text{ }\Omega} = 44 \text{ mA}$$

With I_C approximately equal to 44 mA, the dc collector-ground voltage is

$$V_C = 25 \text{ V} - (44 \text{ mA} \times 150 \text{ }\Omega) = 18.4 \text{ V}$$

EXAMPLE 7-7

Figure 7-9 shows a two-stage amplifier. (A stage is each transistor with its biasing resistors, including R_C and R_E.) Calculate the dc collector-ground voltage for each stage. (Note: *GND* stands for ground.)

112

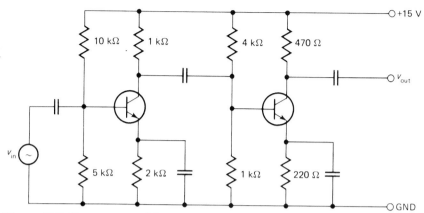

Figure 7-9 Two-stage amplifier.

SOLUTION

The capacitors are open to dc current. Therefore, we can analyze each stage separately because there is no dc interaction between stages. In the first stage, the dc voltage across the 5-kΩ resistor is 5 V. Subtract 0.7 V for the V_{BE} drop and you have 4.3 V across the emitter resistor. Therefore, the dc emitter current in the first stage is

$$I_E = \frac{4.3 \text{ V}}{2 \text{ k}\Omega} = 2.15 \text{ mA}$$

The dc collector-ground voltage is

$$V_C = 15 \text{ V} - (2.15 \text{ mA} \times 1 \text{ k}\Omega) = 12.9 \text{ V}$$

In the second stage, the voltage across the 1-kΩ biasing resistor is 3 V. After subtracting 0.7 V, we have 2.3 V across the emitter resistor. The dc emitter current is

$$I_E = \frac{2.3 \text{ V}}{220 \text{ }\Omega} = 10.5 \text{ mA}$$

and the dc collector-ground voltage is

$$V_C = 15 \text{ V} - (10.5 \text{ mA} \times 470 \text{ }\Omega) = 10.1 \text{ V}$$

113

7-4 EXACT FORMULA FOR VOLTAGE-DIVIDER BIAS

By applying Thevenin's theorem to the base voltage-divider, we can derive the following exact formula for dc emitter current in a voltage-divider bias circuit:

$$I_E = \frac{V_2 - V_{BE}}{R_E + (R_1 \parallel R_2)/\beta_{dc}} \tag{7-12}$$

In the mass-production of transistor circuits, one of the main problems is the variation in β_{dc}. It varies from transistor to transistor of the same type. For instance, a 2N3904 has a minimum β_{dc} of 100 and a maximum β_{dc} of 300. Furthermore, β_{dc} changes with temperature. The variation in β_{dc} makes it virtually impossible to use base bias in linear circuits, because the Q point is unpredictable. But with voltage-divider bias, we can almost eliminate the effect of β_{dc}.

Here's the idea: When we select values for R_1, R_2, and R_E, we can satisfy this condition:

$$R_E \gg \frac{R_1 \parallel R_2}{\beta_{dc}} \tag{7-13}$$

This forces Eq. (7-12) to reduce to

$$I_E = \frac{V_2 - V_{BE}}{R_E} \tag{7-14}$$

Since β_{dc} no longer appears in the equation, the emitter current no longer depends on the value of β_{dc}. Study the following example and you will understand the problem better.

EXAMPLE 7-8

Suppose β_{dc} varies from 100 to 300 in Fig. 7-10a and b. Calculate the minimum and maximum collector current for each circuit.

SOLUTION

In the base-biased circuit of Fig. 7-10a,

$$I_B \cong \frac{30 \text{ V}}{3 \text{ M}\Omega} = 10 \text{ }\mu\text{A}$$

114

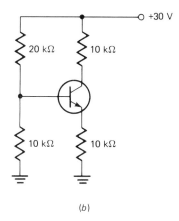

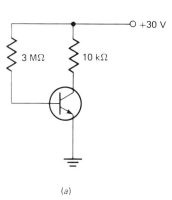

(a)   (b)

Figure 7-10

The minimum collector current is

$$I_C = 100 \times 10 \ \mu A = 1 \ mA$$

and the maximum collector current is

$$I_C = 300 \times 10 \ \mu A = 3 \ mA$$

This huge change in collector current means the Q point will drift all over the load line when transistors are changed or when the temperature varies. Because of this, base bias is worthless in linear circuits.

Voltage-divider bias is different. In Fig. 7-10*b*,

$$R_1 \| R_2 = 20 \ k\Omega \| 10 \ k\Omega = 6.67 \ k\Omega$$

With Eq. (7-12), the minimum emitter current is

$$I_E = \frac{30 \ V - 0.7 \ V}{10 \ k\Omega + 6.67 \ k\Omega/100} = \frac{29.3 \ V}{10,067 \ \Omega} = 2.91 \ mA$$

The maximum emitter current is

$$I_E = \frac{30 \ V - 0.7 \ V}{10 \ k\Omega + 6.67 \ k\Omega/300} = \frac{29.3 \ V}{10,022 \ \Omega} = 2.92 \ mA$$

Even though β_{dc} varies from 100 to 300, the emitter current varies from only 2.91 to 2.92 mA. In other words, the Q point is rock-solid in a voltage-

115

divider bias circuit. This is why voltage-divider bias is used in mass production; it provides a stable Q point for different transistors and changing temperatures. There's no need to use the exact equation when troubleshooting. Assume condition (7-13) is satisfied and use Eq. (7-14) to analyze voltage-divider bias.

7-5 OTHER BIASING CIRCUITS

Up to now, the discussion has been about base bias and voltage-divider bias. Even though voltage-divider bias is the most heavily used, other biasing circuits are occasionally used.

Collector-feedback bias

Figure 7-11a shows *collector-feedback bias*. It offers simplicity (only two resistors) and good low-frequency response (discussed later). The base resistor is returned to the collector rather than to the power supply. This is what distinguishes collector-feedback bias from base bias.

Here's how the *feedback* works. Suppose the temperature increases, causing β_{dc} to increase. This increases the collector current. But as soon as the collector current increases, the collector voltage decreases. This means less voltage drives the base resistor, causing a decrease in base current. Since

$$I_C = \beta_{dc} I_B$$

the decrease in base current partially offsets the increase in β_{dc}. In other words, the collector current will not increase as much as it would in a base-biased circuit.

For your convenience, Fig. 7-11a includes the approximate formulas for $I_{C(sat)}$, I_C, and V_{CE}. If you use this circuit, you can set up a Q point near the middle of the dc load line by selecting a base resistor whose value is

$$R_B = \beta_{dc} R_C \qquad (7\text{-}15)$$

Emitter feedback

Figure 7-11b shows *emitter-feedback bias*. The emitter feedback works like this. When β_{dc} increases, the emitter current increases. This increases the emitter voltage and decreases the voltage across the base resistor. With less voltage across R_B, there is less base current. The decrease in base current

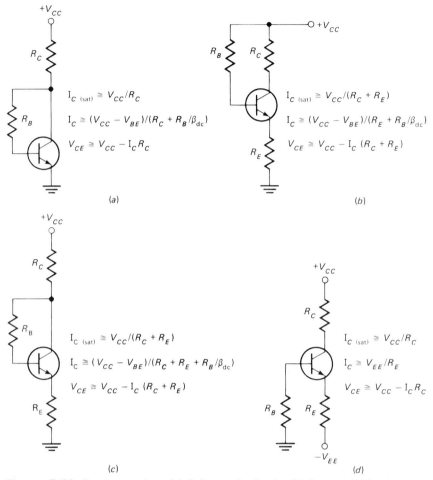

Figure 7-11 Biasing circuits. *(a)* Collector feedback. *(b)* Emitter feedback. *(c)* Collector and emitter feedback. *(d)* Emitter bias.

partially offsets the increase in β_{dc}. In other words, collector current will not increase as much as it would in a base-bias circuit.

Collector and emitter feedback

Figure 7-11c combines collector and emitter feedback. An increase in β_{dc} produces a larger emitter voltage and a smaller collector voltage. Both of these changes reduce the voltage across the base resistor. The decrease in base current partially offsets the increase in β_{dc}.

117

Emitter bias

When positive and negative power supplies are available, *emitter bias* (Fig. 7-11*d*) provides a stable *Q* point. The name "emitter bias" is used because a negative supply forward-biases the emitter diode through resistor R_E.

Here is the key to analyzing the circuit. The emitter-ground voltage is typically less than 1 V. As an approximation, visualize the emitter at ground. Because of this, almost all of the V_{EE} supply voltage appears across R_E, producing an emitter current of

$$I_E \cong \frac{V_{EE}}{R_E} \qquad (7\text{-}16)$$

This says the emitter current is virtually independent of β_{dc}. In other words, emitter bias produces a rock-solid *Q* point, similar to voltage-divider bias.

Emitter bias is extensively used in differential amplifiers, the backbone of linear ICs. A later chapter discusses differential amplifiers and linear ICs. For now, all you have to remember is the key idea. Almost all of V_{EE} is across R_E; therefore, $I_E \cong V_{EE}/R_E$.

Bias stability

A final point: All our biasing formulas neglect thermally produced current and surface-leakage current in the reverse-biased collector diode. This is reasonable, provided the circuit satisfies a certain condition we'll discuss now.

Almost any data sheet lists the value of I_{CBO} at 25°C; this is the current from collector to base with an open emitter. Often, data sheets give the value of I_{CBO} at several temperatures besides 25°C. If not, you can estimate I_{CBO} at other temperatures as follows: I_{CBO} roughly doubles for every 10°C rise.

When I_{CBO} flows through the external base resistance (either R_B or $R_1 \parallel R_2$), it produces a small voltage that our biasing formulas do not take into account. This voltage may shift the *Q* point unless it is negligible compared to the supply voltage. In other words, to have *bias stability* (a solid *Q* point), we have to satisfy this rule of thumb:

$$V_{CC} \gg I_{CBO} R_B$$

at the highest operating temperature. If voltage-divider bias is involved, use $R_1 \parallel R_2$ instead of R_B. For emitter bias, use V_{EE} instead of V_{CC}. With modern silicon transistors, the condition is easily satisfied because I_{CBO} is usually small.

118

7-6 *pnp* BIASING

Figure 7-12*a* is a *pnp* transistor. With the polarities shown for V_{BE} and V_{CE}, the emitter diode is forward-biased and the collector diode is reverse-biased. Conventional current is into the emitter and out of the collector. Electron flow is into the collector and out of the emitter. To use a *pnp* transistor in any of the biasing circuits discussed earlier, replace the positive power supply by a negative power supply (and vice versa for emitter bias).

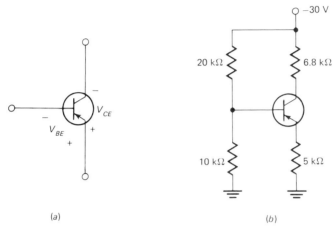

(a)   (b)

Figure 7-12

EXAMPLE 7-9

What is the emitter-ground voltage in Fig. 7-12*b*? The collector-emitter voltage? The collector-ground voltage?

SOLUTION

Because of the voltage divider, the base voltage is -10 V with respect to ground. Allowing 0.7 V for the V_{BE}, the emitter-ground voltage is

$$V_E = -9.3 \text{ V}$$

The emitter current is

$$I_E = \frac{9.3 \text{ V}}{5 \text{ k}\Omega} = 1.86 \text{ mA}$$

The collector-emitter voltage is

119

$$V_{CE} = -30 \text{ V} + (1.86 \text{ mA} \times 11.8 \text{ k}\Omega) = -8.05 \text{ V}$$

The collector-ground voltage is

$$V_C = -30 \text{ V} + (1.86 \text{ mA} \times 6.8 \text{ k}\Omega) = -17.4 \text{ V}$$

7-7 UPSIDE-DOWN BIASING OF *pnp* TRANSISTORS

Ground is a reference point that we can move around. For instance, Fig. 7-13*a* shows voltage-divider bias with a *pnp* transistor. Figure 7-13*b* is the same circuit with the ground removed. Even though the circuit is *floating* (no ground), all transistor currents have the same values as before, because the emitter diode is still forward-biased and the collector diode reverse-biased.

Nothing prevents us from grounding the negative terminal of the power supply to get the circuit of Fig. 7-13*c*. The transistor still operates in the

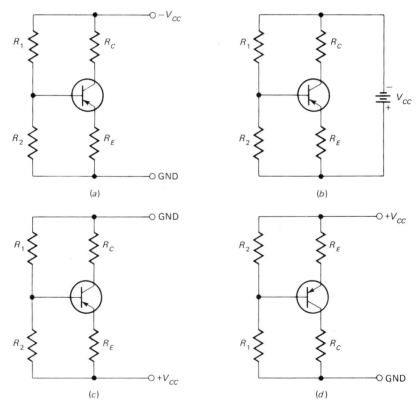

Figure 7-13 Drawing *pnp* transistors upside-down.

120

active region with the same currents as before. It does not matter how a transistor is oriented in space; it works just as well upside down as shown in Fig. 7-13d. In this upside-down biasing circuit, all transistor currents have the same values as when we started (Fig. 7-13a).

Drawing *pnp* transistors upside down with positive power supplies is common drafting practice nowadays. Often, both *npn* and *pnp* transistors are used in the same circuit. By drawing *pnp* transistors upside down, we wind up with a simpler-looking schematic diagram.

EXAMPLE 7-10

What is the dc voltage from each collector to ground in Fig. 7-14?

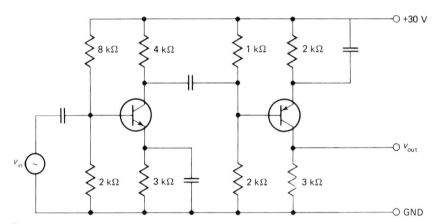

Figure 7-14

SOLUTION

The capacitors are open to dc; therefore, they have nothing to do with the problem. The first stage is voltage-divider biased with an *npn* transistor. Because of the 5:1 voltage divider, 6 V appears across the 2-kΩ base resistor. The emitter current is

$$I_E = \frac{6 \text{ V} - 0.7 \text{ V}}{3 \text{ k}\Omega} = 1.77 \text{ mA}$$

and the collector-ground voltage is

$$V_C = 30 \text{ V} - (1.77 \text{ mA} \times 4 \text{ k}\Omega) = 22.9 \text{ V}$$

The second stage uses a *pnp* transistor with upside-down voltage-divider bias. Because of the 3:1 voltage divider, 10 V appears across the 1-kΩ

121

base resistor. Subtracting 0.7 V for the V_{BE} drop gives 9.3 V across the 2-kΩ emitter resistor. Therefore, the emitter current is

$$I_E = \frac{9.3 \text{ V}}{2 \text{ k}\Omega} = 4.65 \text{ mA}$$

and the collector-ground voltage is

$$V_C = 4.65 \text{ mA} \times 3 \text{ k}\Omega = 13.95 \text{ V}$$

7-8 ANALYZING COMMON-EMITTER AMPLIFIERS

In the typical CE amplifier the Q point is located near the middle of the load line as shown in Fig. 7-15. An ac source drives the base and produces a sinusoidal variation in base and collector current. This forces the instantaneous operating point to swing above and below the Q point. The resulting ac output voltage is usually much larger than the ac input voltage. In other words, we get amplification.

Notice that the transistor operates in the active region throughout the ac cycle. It is never driven into saturation or cutoff. If it were, the peaks of the signal would be clipped off, resulting in *distortion*.

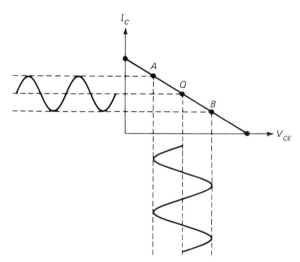

Figure 7-15 AC signal swing on load line.

AC beta

The *ac beta* of a transistor is defined as the ratio of ac collector current to ac base current. In symbols,

$$\beta = \frac{i_c}{i_b} \qquad (7\text{-}17)$$

On data sheets, β is listed as h_{fe}. (Recall $\beta_{dc} = h_{FE}$.) As an example, when the Q point is at 1 mA, a 2N3904 has an h_{fe} from 100 to 400. This means ac collector current is 100 to 400 times greater than ac base current.

Ideal CE transistor

In the common-base connection, the ac signal drives the emitter. The ac resistance of the emitter diode is

$$r_e' \cong \frac{25 \text{ mV}}{I_E}$$

But in the common-emitter connection, the ac signal drives the base. Since base current is much smaller than emitter current, the ac impedance looking into the base terminal is much larger. Because ac collector current approximately equals ac emitter current, ac base current is approximately β times smaller than ac emitter current. For this reason, the ac impedance looking into the base is

$$z_{in(base)} \cong \beta r_e' \qquad (7\text{-}18)$$

Figure 7-16 shows the ac equivalent circuit for a transistor in the CE connection. Notice how the emitter diode is replaced by an ac resistance of $\beta r_e'$ and how the collector diode becomes a current source of βi_b.

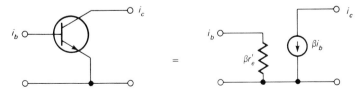

Figure 7-16 Ideal ac equivalent circuit.

Typical CE amplifier

Figure 7-17a is a CE amplifier. Capacitor C_1 couples the ac signal into the base. C_2 couples the amplified signal to a load resistor R_L. Sometimes this load resistance represents another stage. C_3 is called a *bypass* capacitor because it bypasses the emitter to ac ground. In other words, because capacitors appear shorted to the ac signal, the emitter appears grounded in the ac equivalent circuit.

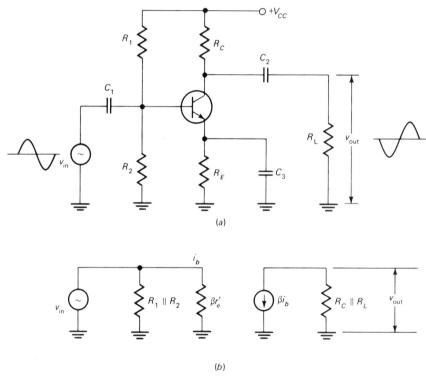

Figure 7-17 CE amplifier. *(a)* Original circuit. *(b)* AC equivalent circuit.

Figure 7-17b shows the ac equivalent circuit. Notice how biasing resistors R_1 and R_2 appear in parallel when the dc supply is reduced to zero. Notice also how R_C is in parallel with R_L; we call this parallel combination the *ac collector resistance,* designated r_C. In symbols,

$$r_C = R_C \parallel R_L \qquad (7\text{-}19)$$

Voltage gain

Since the ac input voltage is across $\beta r_e'$,

$$v_{in} = i_b \beta r_e'$$

The ac output voltage is across the parallel combination of R_C and R_L; therefore,

$$v_{out} = \beta i_b r_C$$

Taking the ratio of v_{out} to v_{in} gives the voltage gain:

$$A = \frac{v_{out}}{v_{in}} = \frac{\beta i_b r_C}{i_b \beta r_e'}$$

or

$$A = \frac{r_C}{r_e'} \qquad (7\text{-}20)$$

Memorize this formula. It is a widely used approximation for the voltage gain of a CE amplifier. It applies to the other biasing circuits as well (base bias, collector-feedback bias, etc., when the emitter is ac grounded).

An alternative form of Eq. (7-20) is

$$A_v = \frac{r_C}{r_e'}$$

where A_v is the voltage gain.

Input impedance

The input impedance of an amplifier is the impedance that the ac source has to drive. In Fig. 7-17b, the ac source has to supply current to the parallel combination of R_1, R_2, and $\beta r_e'$. This means the input impedance is

$$z_{in} = R_1 \| R_2 \| \beta r_e' \qquad (7\text{-}21)$$

Phase inversion

A final point: During the positive half cycle of ac input voltage, the base current increases, causing the collector current to increase. This produces a greater drop across the collector resistor, which means collector voltage

Transistor Circuit Approximations

decreases. On the negative half cycle of ac input voltage, collector voltage increases. For this reason, the ac output signal in any CE amplifier is always 180° out of phase with the ac input signal (see Fig. 7-17*a*).

EXAMPLE 7-11

The 2N3904 of Fig. 7-18 has a β of 200. What is the ac output voltage? The input impedance?

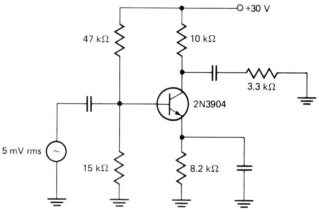

Figure 7-18 Swamping resistor.

SOLUTION

The dc voltage across the 15-kΩ biasing resistor is

$$V_2 = \frac{15 \text{ k}\Omega}{62 \text{ k}\Omega} 30 \text{ V} = 7.26 \text{ V}$$

The dc voltage across the emitter resistor is

$$V_E = 7.26 \text{ V} - 0.7 \text{ V} = 6.56 \text{ V}$$

Therefore, the dc emitter current is

$$I_E = \frac{6.56 \text{ V}}{8.2 \text{ k}\Omega} = 0.8 \text{ mA}$$

and the ac resistance of the emitter diode is

$$r'_e = \frac{25 \text{ mV}}{0.8 \text{ mA}} = 31.3 \text{ }\Omega$$

126

Now, we can calculate the voltage gain:

$$A = \frac{r_C}{r_e'} = \frac{10 \text{ k}\Omega \,\|\, 3.3 \text{ k}\Omega}{31.3 \ \Omega} = 79.3$$

The ac output voltage is

$$v_{\text{out}} = A v_{\text{in}} = 79.3 \times 5 \text{ mV} = 0.397 \text{ V}$$

The ac resistance looking into the base is

$$z_{\text{in(base)}} = \beta r_e' = 200 \times 31.3 \ \Omega = 6.26 \text{ k}\Omega$$

Finally, the input impedance is

$$z_{\text{in}} = R_1 \,\|\, R_2 \,\|\, \beta r_e' = 47 \text{ k}\Omega \,\|\, 15 \text{ k}\Omega \,\|\, 6.26 \text{ k}\Omega = 4.04 \text{ k}\Omega$$

7-9 SWAMPING THE EMITTER DIODE

We have been using the following formula for the ac resistance of the emitter diode:

$$r_e' = \frac{25 \text{ mV}}{I_E}$$

This is an ideal formula for an abrupt pn junction at 25°C. When the junction is not abrupt or when the temperature changes, r_e' will change. Any change in r_e' changes the voltage gain of a CE amplifier. In some applications, such a change in voltage gain is not acceptable.

Swamping resistor

To desensitize an amplifier from the changes in r_e', we can *swamp* the emitter diode by adding a resistor between the emitter and ac ground, as shown in Fig. 7-19. This produces emitter feedback, reducing the effect of r_e'. Since r_E is in series with r_e' in the ac equivalent circuit, the voltage gain is

$$A = \frac{r_C}{r_E + r_e'} \qquad \text{(7-22)}$$

The ac impedance looking into the base is

$$z_{\text{in(base)}} = \beta(r_E + r_e') \qquad \text{(7-23)}$$

127

Transistor Circuit Approximations

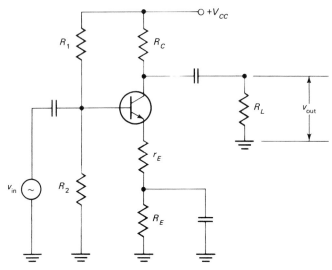

Figure 7-19

and the input impedance of the stage is

$$z_{in} = R_1 \| R_2 \| z_{in(base)} \qquad \text{(7-24)}$$

Heavy swamping

When r_E is much greater than r'_e, the value of r'_e has almost no effect on voltage gain and input impedance. In fact, with heavy swamping $(r_E \gg r'_e)$, Eqs. (7-22) and (7-23) simplify to

$$A \cong \frac{r_C}{r_E} \qquad \text{(7-25)}$$

and
$$z_{in(base)} \cong \beta r_E$$

Because r_E is a fixed resistor, the voltage gain is fixed. This means we can change transistors and still have the same value of voltage gain. Likewise, changes in temperature can no longer affect the voltage gain because r'_e has been swamped out.

EXAMPLE 7-12

In Fig. 7-20, r'_e varies from 50 to 75 Ω. What are the minimum and maximum voltage gains?

128

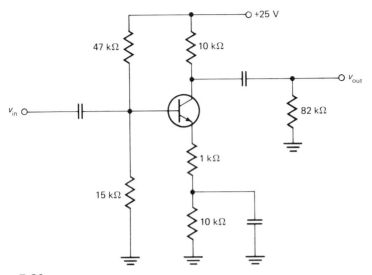

Figure 7-20

SOLUTION

The minimum gain is

$$A = \frac{r_C}{r_E + r_e'} = \frac{10 \text{ k}\Omega \parallel 82 \text{ k}\Omega}{1 \text{ k}\Omega + 75 \text{ }\Omega} = 8.29$$

The maximum gain is

$$A = \frac{10 \text{ k}\Omega \parallel 82 \text{ k}\Omega}{1 \text{ k}\Omega + 50 \text{ }\Omega} = 8.49$$

Even though r_e' increases from 50 to 75 Ω (a 50-percent change), A only decreases from 8.49 to 8.29 (a 2.36-percent change).

The price we pay for swamping out r_e' is a loss of gain. But this is no problem. If you need more gain, use two stages like the swamped amplifier of Fig. 7-20. Two stages with a gain of 8.29 have an overall gain of

$$A = A_1 \times A_2 = 8.29 \times 8.29 = 68.7$$

EXAMPLE 7-13

What is the input impedance if $\beta = 200$ in Fig. 7-20? (Use $r_e' = 50$ Ω).

SOLUTION

$$z_{\text{in(base)}} = \beta(r_E + r_e') = 200(1.05 \text{ k}\Omega) = 210 \text{ k}\Omega$$

129

Transistor Circuit Approximations

The input impedance is

$$z_{\text{in}} = R_1 \parallel R_2 \parallel z_{\text{in(base)}} = 47 \text{ k}\Omega \parallel 15 \text{ k}\Omega \parallel 210 \text{ k}\Omega = 10.8 \text{ k}\Omega$$

GLOSSARY

active region The area where the collector curves are almost horizontal. In the active region the emitter diode is forward-biased and the collector diode is reverse-biased.

amplification Increasing the amplitude of a signal.

bypass capacitor One used to ac-ground a point.

cutoff The lower end of the load line. At this point the emitter diode comes out of forward bias and the collector current goes to approximately zero.

linear circuit One designed to operate in the active region, usually to amplify signals. Linear circuits are never driven into saturation or cutoff.

load line A graph of all possible operating points.

phase inversion Refers to the output signal being 180° out of phase with the input signal.

saturation The upper end of the load line. At this point the collector diode comes out of reverse bias and no further increase in collector current is possible.

SELF-TESTING REVIEW

Read each of the following and provide the missing words. Answers appear at the beginning of the next question.

1. The almost vertical region of collector curves is called the _____ region. Too much collector-emitter voltage will break down the collector diode. The _____ region is where the collector curves are almost horizontal.

2. *(saturation, active)* The dc load line is a graph of all possible dc operating points. The upper end of this line is called the _____ point; the lower end is the _____ point. The Q point lies somewhere on the dc load line.

3. *(saturation, cutoff)* Linear transistor circuits always operate in the _____ region. The most widely used bias for linear transistor circuits

130

is _____ bias. With this kind of bias, variations in β_{dc} have almost no effect.

4. *(active, voltage-divider)* When positive power supplies are being used, *pnp* transistors are usually drawn _____. Often, both *npn* and *pnp* transistors are in the same circuit. By drawing *pnp* transistors _____, we wind up with a simpler-looking schematic diagram.

5. *(upside down, upside down)* The ac beta is the ratio of ac collector current to ac _____ current. The ac impedance looking into the base of a CE transistor is _____ times r'_e. The reason is that ac base current is approximately β times smaller than ac emitter current.

6. *(base, β)* The voltage gain of a CE amplifier equals _____ divided by r'_e. This applies to the other biasing circuits like base bias, collector-feedback bias, etc. The input _____ is the equivalent parallel resistance of the base-biasing resistors and $z_{in(base)}$.

7. *(r_C, impedance)* To reduce the effect of r'_e, we can swamp the emitter diode by adding a _____ between the emitter and ac ground. The voltage gain of a heavily swamped amplifier equals r_C divided by r_E. Because these resistances are fixed, the voltage gain is _____.

8. *(resistor, fixed)* The input impedance of a swamped amplifier is higher than a CE amplifier because r_E is added to r'_e. This sum multiplied by β gives the ac impedance looking into the base.

PROBLEMS

7-1. $R_B = 2.2$ MΩ and $R_C = 6.8$ kΩ in Fig. 7-21a. If $h_{FE} = 175$, what does I_C equal? And what does V_{CE} equal?

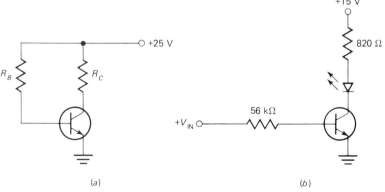

(a)   (b)

Figure 7-21

Transistor Circuit Approximations

7-2. If $\beta_{dc} = 200$, $R_B = 1.5$ MΩ, and $R_C = 4.7$ kΩ in Fig. 7-21a, what do I_C and V_C equal?

7-3. In Fig. 7-21a, $R_C = 5.1$ kΩ and $\beta_{dc} = 250$. What does $I_{C(sat)}$ equal? What is the value of R_B that just saturates the transistor?

7-4. $R_C = 2.2$ kΩ and $h_{FE} = 150$ in Fig. 7-21a. What is the value of R_B that locates the Q point near the middle of the dc load line?

7-5. The LED of Fig. 7-21b has 1.8 V across it when the transistor is saturated. How much current is there through the LED? If $\beta_{dc} = 100$, what is the minimum value of V_{IN} that just produces saturation?

7-6. What is the collector current in Fig. 7-22a? The collector-ground voltage? The collector-emitter voltage?

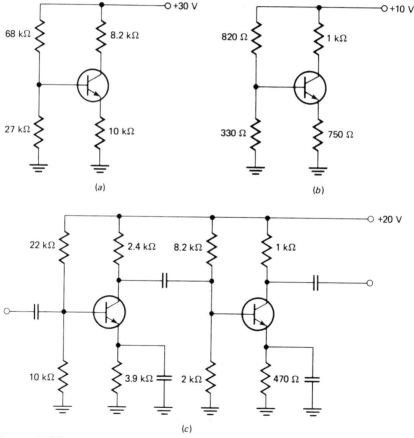

(a)

(b)

(c)

Figure 7-22

132

7-7. In Fig. 7-22b, what does I_C equal? V_C? V_E?

7-8. Figure 7-22c shows a two-stage amplifier.   What do I_C and V_C equal for each stage?

7-9. If $\beta_{dc} = 150$, what are the values of I_C and V_C in Fig. 7-23a?

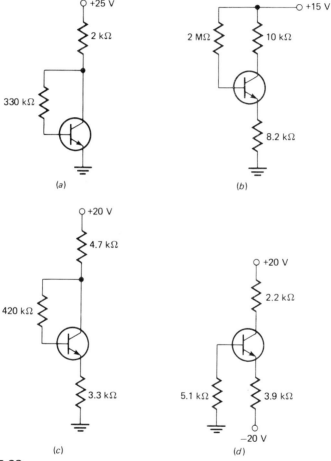

(a)

(b)

(c)

(d)

Figure 7-23

7-10. In Fig. 7-23b, what does I_C equal for $\beta_{dc} = 100$?   What are values of V_C and V_E?

7-11. If $h_{FE} = 120$ in Fig. 7-23c, what does the collector current equal? The collector-ground voltage?   The emitter-ground voltage?

Transistor Circuit Approximations

7-12. In Fig. 7-23*d*, what does I_E equal? V_C?

7-13. Calculate the values of I_C and V_C in Fig. 7-24*a*.

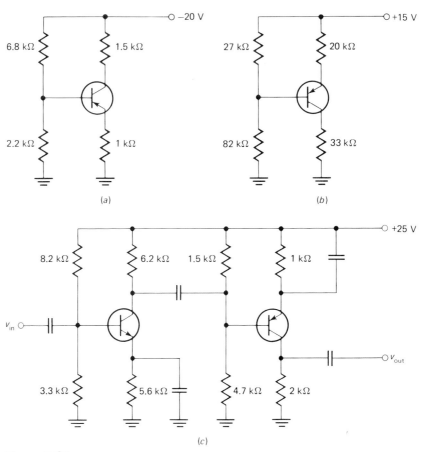

(a)

(b)

(c)

Figure 7-24

7-14. Find I_C and V_C in Fig. 7-24*b*.

7-15. Work out the collector-to-ground voltage for each stage of Fig. 7-24*c*.

7-16. What is the voltage gain in Fig. 7-25*a*?

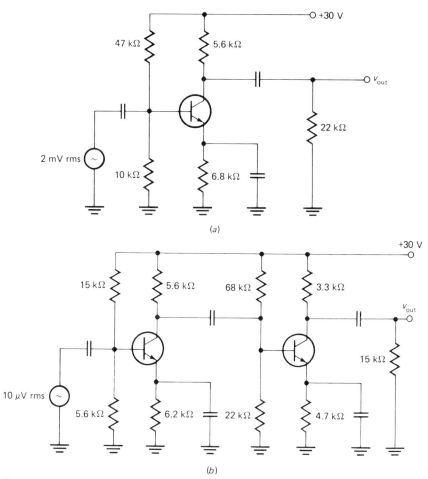

(a)

(b)

Figure 7-25

7-17. If $\beta = 250$ in Fig. 7-25a, what is the input impedance? The ac output voltage?

7-18. Each stage of Fig. 7-25b has an h_{fe} of 200. What does the ac output voltage equal?

Transistor Circuit Approximations

7-19. Each transistor of Fig. 7-26 has a β of 300. If r'_e is negligible, what is the ac output voltage for an ac input voltage of 575 μV rms?

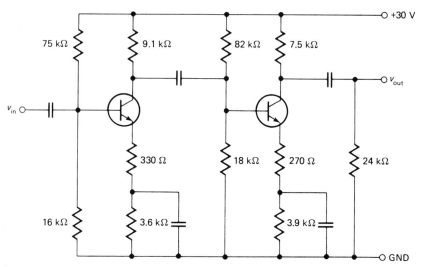

Figure 7-26

7-20. In Fig. 7-26, each transistor has a β of 250. Including r'_e in your calculations, what does v_{out} equal when v_{in} equals 1 mV rms?

Chapter 8

Common-Collector Approximations

We have studied two basic transistor connections, the common base (CB) and the common emitter (CE). This chapter is about the third basic connection, known as the *common collector* (CC). The main advantage of the CC connection is its high input impedance.

8-1 VOLTAGE GAIN

Figure 8-1 shows a CC stage. The collector is ac-grounded through the power supply. An ac input signal v_{in} is coupled into the base. This produces sinusoidal variations in emitter current, and a sinusoidal voltage appears across R_E. The ac signal is then coupled to the load resistor R_L.

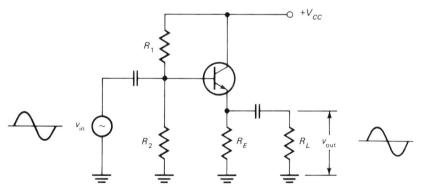

Figure 8-1 Emitter follower.

Figure 8-2 is the ac equivalent circuit. Resistance r_E is the equivalent parallel resistance of R_E and R_L:

Transistor Circuit Approximations

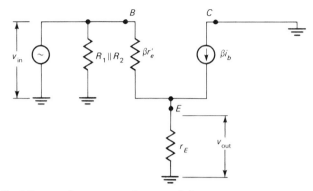

Figure 8-2 AC equivalent circuit of emitter follower.

$$r_E = R_E \parallel R_L$$

The ac input signal v_{in} produces an ac base current i_b through $\beta r_e'$ and an ac emitter current i_e through r_E. Therefore, the ac input voltage equals

$$v_{in} = i_b \beta r_e' + i_e r_E$$

The ac output voltage is across r_E and equals

$$v_{out} = i_e r_E$$

The ratio of v_{out} to v_{in} is

$$A = \frac{v_{out}}{v_{in}} = \frac{i_e r_E}{i_b \beta r_e' + i_e r_E} \cong \frac{i_e r_E}{i_e r_e' + i_e r_E}$$

which reduces to

$$A \cong \frac{r_E}{r_E + r_e'} \tag{8-1}$$

When r_E is much greater than r_e', A is approximately unity. Equation (8-1) often appears as $A_v \cong r_E/(r_E + r_e')$.

In other words, the voltage gain of a typical CC stage approaches unity. This means the output signal has approximately the same amplitude and phase as the input signal. This is why a CC stage is called an *emitter follower;* the signal on the emitter follows the signal on the base, duplicating both amplitude and phase.

138

EXAMPLE 8-1

If $I_E = 1$ mA, $R_E = 1$ kΩ, and $R_L = 5.6$ kΩ, what is the voltage gain of an emitter follower?

SOLUTION

First, get the value of r_e':

$$r_e' \cong \frac{25 \text{ mV}}{1 \text{ mA}} = 25 \text{ }\Omega$$

Next, determine r_E:

$$r_E = R_E \parallel R_L = 1 \text{ k}\Omega \parallel 5.6 \text{ k}\Omega = 848 \text{ }\Omega$$

Finally, calculate the voltage gain:

$$A \cong \frac{r_E}{r_E + r_e'} = \frac{848}{848 + 25} = 0.971$$

8-2 INPUT IMPEDANCE

Why use an emitter follower if the voltage gain is less than unity? Because an emitter follower can step up impedance, similar to a transformer. In Fig. 8-2, the ac impedance looking into the base is

$$z_{\text{in(base)}} = \frac{v_{\text{in}}}{i_b} = \frac{i_b \beta r_e' + i_e r_E}{i_b}$$

Since $i_c \cong i_e$, the foregoing reduces to

$$z_{\text{in(base)}} \cong \beta(r_E + r_e') \qquad (8\text{-}2)$$

When r_E is much greater than r_e', $z_{\text{in(base)}}$ approximately equals βr_E.

Equation (8-2) says the ac load impedance is stepped up by a factor of β. This is the whole point of using an emitter follower. If we want to increase the ac impedance of a load, we can use an emitter follower in front of this load.

EXAMPLE 8-2

In Fig. 8-3, what is the input impedance of the first stage if each transistor has a β of 100?

139

Transistor Circuit Approximations

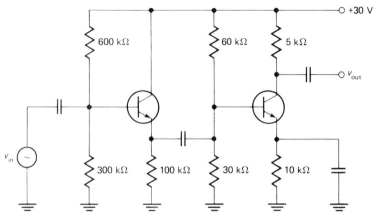

Figure 8-3 Source impedance and input impedance.

SOLUTION

In the second stage, 10 V appear across the 30-kΩ biasing resistor. Ignoring the V_{BE} drop, there is approximately 1 mA of dc emitter current and r_e' is 25 Ω. Therefore,

$$z_{in(base)} = \beta r_e' = 100 \times 25 \ \Omega = 2.5 \ k\Omega$$

The input impedance of the second stage includes the biasing resistors:

$$z_{in} = R_1 \parallel R_2 \parallel \beta r_e' = 60 \ k\Omega \parallel 30 \ k\Omega \parallel 2.5 \ k\Omega = 2.22 \ k\Omega$$

This is the load resistance R_L seen by the first stage.

In the first stage, the ac resistance seen by the emitter is

$$r_E = R_E \parallel R_L = 100 \ k\Omega \parallel 2.22 \ k\Omega = 2.17 \ k\Omega$$

Ignoring the V_{BE} drop, the first stage has a dc emitter current of approximately 0.1 mA and an r_e' of 250 Ω. As a result, the ac impedance looking into the base of the first stage is

$$z_{in(base)} = \beta(r_E + r_e') = 100(2170 \ \Omega + 250 \ \Omega) = 242 \ k\Omega$$

The input impedance of the first stage includes the biasing resistors:

$$z_{in} = R_1 \parallel R_2 \parallel z_{in(base)} = 600 \ k\Omega \parallel 300 \ k\Omega \parallel 242 \ k\Omega = 110 \ k\Omega$$

The point is this: If the ac source had to drive the second stage directly, it would see an input impedance of 2.22 kΩ. Because of the emitter follower, however, the ac source looks into an impedance of 110 kΩ. This higher input impedance means there's less loading of the source.

8-3 POWER GAIN

We do not get voltage gain with an emitter follower, but we do get *power gain*. The ac output power of an emitter follower is

$$p_e = i_e^2 r_E$$

and the ac input power to the base is

$$p_b = i_b^2 z_{\text{in(base)}} = i_b^2 \beta (r_E + r_e')$$

The power gain (symbolized by G) is the ratio of p_e to p_b:

$$G = \frac{p_e}{p_b} = \frac{i_e^2 r_E}{i_b^2 \beta (r_E + r_e')}$$

Since $i_c \cong i_e$, this reduces to

$$G = \beta \frac{r_E}{r_E + r_e'} \qquad \text{(8-3)}$$

In this formula, the first factor β is called the *current gain,* and the second factor is the voltage gain. The product of current gain and voltage gain is the power gain.

For the case of r_E being much greater than r_e', the power gain approximately equals β. In other words, the typical emitter follower has a power gain approaching β.

EXAMPLE 8-3

What is the power gain for the first stage of Fig. 8-3?

SOLUTION

In Example 8-2, we found $r_E = 2.17$ kΩ and $r_e' = 250$ Ω for the first stage. The voltage gain is

$$A = \frac{r_E}{r_E + r_e'} = \frac{2170\ \Omega}{2170\ \Omega + 250\ \Omega} = 0.897$$

141

Transistor Circuit Approximations

The power gain is

$$G = \beta \frac{r_E}{r_E + r'_e} = 100 \times 0.897 = 89.7$$

8-4 SOURCE IMPEDANCE

Every ac source has some internal impedance (equivalent to its Thevenin impedance). This source impedance is too large to ignore when the load impedance is small.

Figure 8-4 shows an ac source with an internal impedance R_S. Alternating current from the source has to pass through R_S to reach the base of the

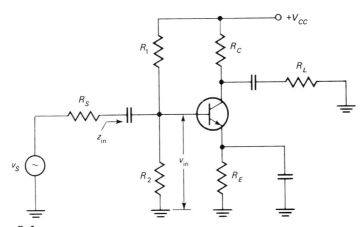

Figure 8-4

CE amplifier. Because of the drop across R_S, the input voltage v_{in} may be considerably less than the source voltage v_S. In fact, voltage-divider action (z_{in} in series with R_S) gives us an ac input voltage of

$$v_{in} = \frac{z_{in}}{R_S + z_{in}} v_S \tag{8-4}$$

When z_{in} is much greater than R_S, v_{in} approximately equals v_S. This means almost all the ac source voltage reaches the base when the source is lightly loaded.

EXAMPLE 8-4

What is the ac input voltage in Fig. 8-5?

142

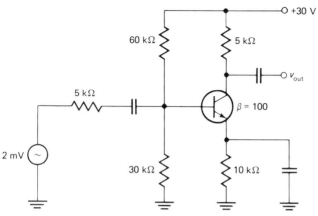

Figure 8-5

SOLUTION

Ignoring the V_{BE} drop, I_E is approximately 1 mA and r'_e is 25 Ω. With $\beta = 100$, $z_{in(base)} = 2.5$ kΩ. The input impedance of the stage is

$$z_{in} = R_1 \,\|\, R_2 \,\|\, z_{in(base)} = 60 \text{ k}\Omega \,\|\, 30 \text{ k}\Omega \,\|\, 2.5 \text{ k}\Omega = 2.22 \text{ k}\Omega$$

The ac input voltage is

$$v_{in} = \frac{z_{in}}{R_S + z_{in}} \, v_s = \frac{2.22 \text{ k}\Omega}{5 \text{ k}\Omega + 2.22 \text{ k}\Omega} \, 2 \text{ mV} = 0.615 \text{ mV}$$

As you see, less than one-third of the ac source voltage reaches the base.

8-5 USING AN EMITTER FOLLOWER

The main application of an emitter follower is to avoid the loss of ac voltage across the source resistance. Whenever z_{in} is smaller than R_S, we can insert an emitter follower to step up the impedance level. This reduces the loading on the source and allows more of the source voltage to reach the base of the amplifier.

EXAMPLE 8-5

Both transistors of Fig. 8-6 have a β of 100. What is the ac input voltage to the base of the first stage? The ac output voltage from the second stage?

143

Transistor Circuit Approximations

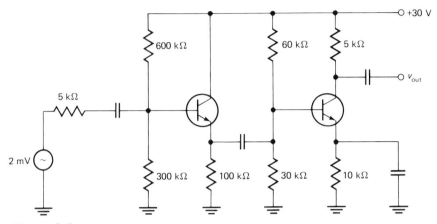

Figure 8-6

SOLUTION

In Example 8-2, we analyzed this two-stage amplifier and found the input impedance of the first stage was

$$z_{in} = 110 \text{ k}\Omega$$

In Fig. 8-6, the source impedance is 5 kΩ; therefore, the ac input voltage to the first stage is

$$v_{in} = \frac{z_{in}}{R_S + z_{in}} v_S = \frac{110 \text{ k}\Omega}{5 \text{ k}\Omega + 110 \text{ k}\Omega} 2 \text{ mV} = 1.91 \text{ mV}$$

In Example 8-3, we worked out the voltage gain of the first stage:

$$A_1 = 0.897$$

In Fig. 8-6, the second stage has a voltage gain of

$$A_2 = \frac{r_C}{r'_e} = \frac{5 \text{ k}\Omega}{25 \text{ }\Omega} = 200$$

The overall voltage gain of the two stages is

$$A = A_1 A_2 = 0.897 \times 200 = 179$$

So, the ac output voltage from the second stage is

$$v_{out} = A v_{in} = 179 \times 1.91 \text{ mV} = 342 \text{ mV}$$

144

Compare Fig. 8-5 with Fig. 8-6. Without the emitter follower, more than two-thirds of the ac source voltage is lost across the source impedance in Fig. 8-5. But with the emitter follower (Fig. 8-6), almost all the ac source voltage reaches the base of the first stage. Since the voltage gain of the emitter follower approaches unity, the final ac output voltage is much larger in Fig. 8-6 than in Fig. 8-5.

Using an emitter follower is often called *buffering* (to buffer means to isolate). The emitter follower isolates the low input impedance of the CE stage from the source impedance. It does this by stepping up the impedance to the point where the ac source is no longer heavily loaded.

8-6 THE DARLINGTON PAIR

The higher the β, the higher the input impedance of an emitter follower. One way to increase β is with a *Darlington pair,* two transistors connected as shown in Fig. 8-7a. The ac emitter current of the first transistor drives

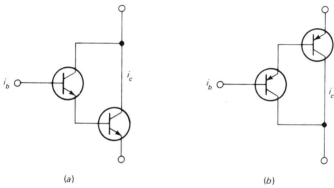

(a)   (b)

Figure 8-7 *(a) npn* Darlington. *(b) pnp* Darlington.

the base of the second transistor. Because of this, the ac collector current in the second transistor is

$$i_c \cong \beta_1 \beta_2 i_b$$

In other words, a Darlington pair has an overall beta equal to the product of the individual betas:

$$\beta = \beta_1 \beta_2 \qquad \text{(8-5)}$$

As an example, if $\beta_1 = 100$ and $\beta_2 = 50$, the Darlington pair has an effective beta of

145

Transistor Circuit Approximations

$$\beta = 100 \times 50 = 5000$$

Figure 8-7*b* is a *pnp* Darlington pair. This too has an overall beta equal to the product of the individual betas.

Transistor manufacturers can put a Darlington pair inside a single transistor package. This three-terminal device acts like a single transistor with an extremely high beta. For instance, a 2N2785 is an *npn* Darlington transistor with a minimum β of 2000 and a maximum β of 20,000.

The high β of a Darlington emitter follower often produces a $z_{in(base)}$ of more than a megohm. In this case, we can no longer neglect the reverse current of the collector diode and must use this formula for the input impedance of the first base:

$$z_{in(base)} = \beta(r_E + r_e') \| r_c'$$

where r_c' is the ac resistance of the input collector diode. This ac resistance is typically in megohms. On data sheets, $r_c' \cong h_{ob}$.

EXAMPLE 8-6

The Darlington pair of Fig. 8-8 has an h_{fe} of 5000, and the input transistor has an h_{ob} of 10 MΩ. Ignoring r_e', what does $z_{in(base)}$ equal? The z_{in} of the stage?

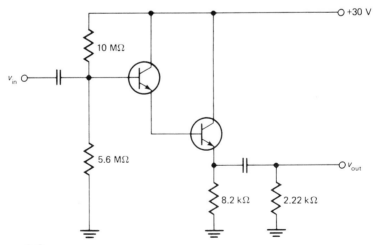

Figure 8-8

SOLUTION

The load resistance seen by the output emitter is

146

$$r_E = 8.2 \text{ k}\Omega \parallel 2.22 \text{ k}\Omega = 1.75 \text{ k}\Omega$$

The ac impedance looking into the first base is

$$z_{\text{in(base)}} = \beta(r_E + r'_e) \parallel r'_c = 5000(1.75 \text{ k}\Omega) \parallel 10 \text{ M}\Omega = 4.67 \text{ M}\Omega$$

The input impedance of the stage is

$$z_{\text{in}} = R_1 \parallel R_2 \parallel z_{\text{in(base)}} = 10 \text{ M}\Omega \parallel 5.6 \text{ M}\Omega \parallel 4.67 \text{ M}\Omega = 2.03 \text{ M}\Omega$$

8-7 THE ZENER REGULATOR

As you may recall, the output of a zener regulator changes slightly when the zener current changes. The change in output voltage is

$$\Delta V_{\text{OUT}} = \Delta I_Z r_Z$$

or

$$\Delta V_{\text{OUT}} = -\Delta I_L r_Z$$

Figure 8-9 shows how to reduce the changes in output voltage. Resistor R_S and the zener diode are a zener regulator as previously described. The regulated zener voltage then drives an emitter follower, and the output of

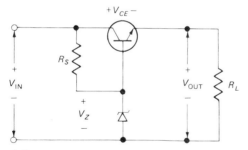

Figure 8-9 Zener follower.

the emitter follower goes to the load resistor R_L. Because of the current gain of the emitter follower, any change in load current is reduced by a factor of β when viewed from the base. The formula for changes in the output voltage is

$$\Delta V_{\text{OUT}} = \frac{-\Delta I_L}{\beta} r_Z \tag{8-6}$$

147

Because of this, the voltage regulation is improved by a factor of β.

Using an emitter follower in cascade with a zener regulator also increases the maximum load current the regulator can handle. Without the emitter follower, the zener diode can regulate from tens to hundreds of milliamperes, depending on the diode type. But with the emitter follower and its current gain, load currents can be in the ampere range. When necessary, a Darlington pair can be used to further increase the load-current capability.

Figure 8-9 is an example of a *series* voltage regulator. The collector-emitter terminals are in series with the load. Because of this, load current must pass through the transistor, and this is the reason the transistor is often called a *pass transistor*. The voltage across the pass transistor is

$$V_{CE} = V_{IN} - V_{OUT}$$

and its power dissipation is

$$P_D = (V_{IN} - V_{OUT}) I_L \qquad \textbf{(8-7)}$$

GLOSSARY

buffer A device or circuit used to isolate two other circuits or stages. The emitter follower is an example of a buffer.

common collector One of the three basic ways to connect a transistor. The input signal is applied to the base, and the output signal is taken from the emitter.

current gain The ratio of output current to input current. For the emitter follower, the current gain equals β or h_{fe}.

power gain The ratio of output power to input power.

source impedance The same as the Thevenin impedance of the source. For audio systems (20 to 20,000 Hz), source impedances are usually 600 Ω. For microwave and other high-frequency systems, source impedance is typically 50 Ω.

voltage gain The ratio of output voltage to input voltage. With an emitter follower, the voltage gain approaches unity with r_e' is negligible.

SELF-TESTING REVIEW

Read each of the following and provide the missing words. Answers appear at the beginning of the next question.

148

1. Two basic transistor connections are the common base and the common emitter. The third basic connection is called the common _____. The voltage gain of this connection is always less than _____. When r_E is much greater than r_e', the voltage gain approaches _____.

2. *(collector, unity, unity)* Because the output signal has approximately the same amplitude and phase as the input signal, a CC stage is often called an emitter _____.

3. *(follower)* The emitter follower steps up the impedance by a factor of _____. Although the voltage gain is less than unity, the emitter follower has a power _____. It equals the product of current gain and voltage gain.

4. *(β, gain)* Every ac source has some internal impedance. Since ac input current passes through this internal impedance, we can get a voltage _____ across this impedance. This means less signal arrives at the input to the stage.

5. *(drop)* The main use of an emitter follower is to avoid signal loss across the ac source impedance. Whenever z_{in} is small, we can insert an emitter follower to step up the _____ by a factor of _____. Because of its near-unity gain, the emitter follower can drive the next stage with almost the entire source voltage.

6. *(impedance, β)* Using an emitter follower is referred to as buffering. The emitter follower can isolate a _____ impedance from a high source impedance. It does this by stepping up the impedance by a factor of β.

7. *(low)* The Darlington pair is a connection of _____ transistors where the emitter of the first drives the base of the second. Because of this, the overall beta equals the _____ of the individual betas.

8. *(two, product)* The zener follower is a zener regulator driving an emitter follower. Because of the current gain of the emitter follower, regulation is improved by a factor of β. The maximum load-current capability is also increased.

PROBLEMS

8-1. The load on an emitter follower is $r_E = 1$ kΩ. If $r_e' = 35$ Ω, what is the voltage gain?

8-2. In an emitter follower, $I_E = 0.5$ mA, $R_E = 2.2$ kΩ, and $R_L = 8.2$ kΩ. What is the voltage gain?

149

Transistor Circuit Approximations

8-3. What is the voltage gain in Fig. 8-10*a*? The ac impedance looking into the base? The input impedance of the stage?

8-4. Figure 8-10*b* shows a *pnp* emitter follower. What are the voltage gain and the input impedance?

8-5. What is the power gain in Fig. 8-10*a*? Fig. 8-10*b*?

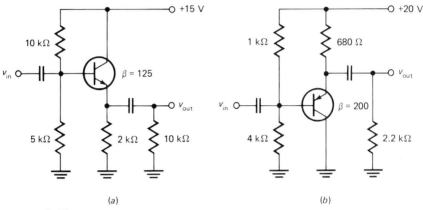

(a)   (b)

Figure 8-10

8-6. In Fig. 8-11, all transistors have a β of 150. If $v_{in} = 1$ mV, what does v_{out} equal? What is the input impedance of the first stage? (Ignore r'_e and r'_c in the first stage.)

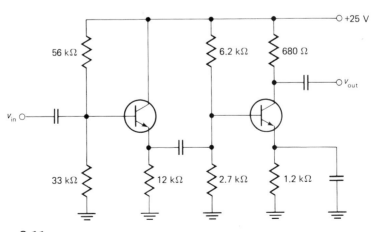

Figure 8-11

150

8-7. If all transistors of Fig. 8-12 have a β of 300, what is the overall voltage gain? The input impedance of the first stage? (Ignore r'_e and r'_c in the first stage.)

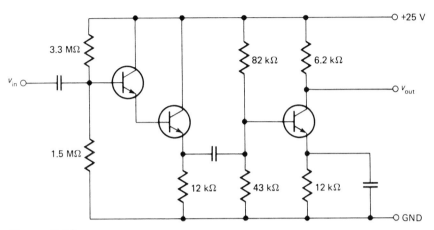

Figure 8-12

8-8. In Fig. 8-13, $V_Z = 7.5$ V, $V_{BE} = 0.7$ V, and $\beta_{dc} = 100$. What is the load voltage? The base current? The zener current?

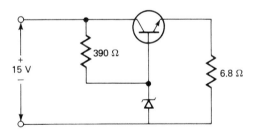

Figure 8-13

Chapter 9

Class A
Power Amplifiers

The end stages of many systems are large-signal amplifiers, where the emphasis is on power gain rather than voltage gain. Amplifiers like this are called *power amplifiers.* Transistors used in small-signal amplifiers are referred to as *small-signal transistors;* those in power amplifiers are *power transistors.* As a guide, a small-signal transistor has a power dissipation of less than 1 W; a power transistor has more than 1 W.

9-1 THE AC LOAD LINE OF A CE AMPLIFIER

Every amplifier sees two loads: a dc load and an ac load. To find the saturation and cutoff points on the dc load line, you have to analyze the dc equivalent circuit. To find the saturation and cutoff points on the ac load line, you have to analyze the ac equivalent circuit.

The Q point

The quiescent collector current and voltage are the I_C and V_{CE} when there is no input signal. No matter what kind of bias is used, the procedure for locating the Q point is the same. You calculate I_C and V_{CE} using the dc equivalent circuit. Then you plot the Q point with I_C on the vertical axis and V_{CE} on the horizontal axis. To distinguish the Q point in our discussions, we will use I_{CQ} for the quiescent collector current and V_{CEQ} for the quiescent collector-emitter voltage.

Saturation and cutoff

Figure 9-1*a* shows the ac equivalent circuit for a CE amplifier. The emitter is at ac ground and the collector drives an ac resistance of r_C. When the

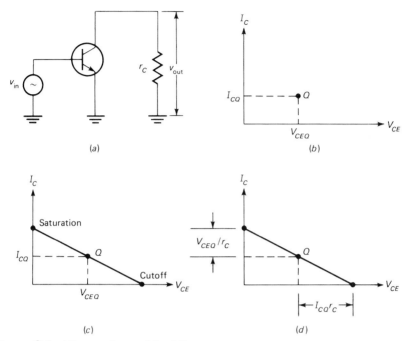

Figure 9-1 AC operation and load line.

ac input signal is zero, the transistor operates at the Q point shown in Fig. 9-1b. When an ac input signal is used, however, the instantaneous operating point swings above and below the Q point, tracing out the ac load line of Fig. 9-1c. If the ac signal is large enough, the transistor is driven into saturation and cutoff.

When the transistor is saturated, V_{CE} is approximately zero. Because of this, the total voltage change from the Q point to saturation is V_{CEQ}, and the change in collector current is

$$\Delta I_{CQ} = \frac{V_{CEQ}}{r_C} \qquad (9\text{-}1)$$

When the transistor is cut off, I_C is approximately zero. As a result, the total current change from the Q point to cutoff is I_{CQ}, and the change in collector-emitter voltage is

$$\Delta V_{CEQ} = I_{CQ} r_C \qquad (9\text{-}2)$$

Figure 9-1d shows these changes. When the operating point swings from

153

Transistor Circuit Approximations

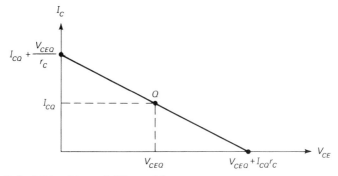

Figure 9-2 AC load line of CE amplifier.

the Q point to the saturation point, the change in collector current equals V_{CEQ}/r_C. On the other hand, when the operating point swings from the Q point to the cutoff point, the maximum change in voltage is $I_{CQ}r_C$.

Figure 9-2 summarizes the ac load line for any CE amplifier. As you see, the quiescent point has coordinates of I_{CQ} and V_{CEQ}. The saturation value of collector current is

$$I_{C(\text{sat})} = I_{CQ} + \frac{V_{CEQ}}{r_C} \qquad \textbf{(9-3)}$$

The maximum collector-emitter voltage occurs at cutoff and equals

$$V_{CE(\text{cutoff})} = V_{CEQ} + I_{CQ}r_C \qquad \textbf{(9-4)}$$

Maximum ratings

The breakdown voltage of the collector diode depends on whether the base is driven by a voltage source or a current source. If a voltage source is involved, the base appears shorted to the emitter when the transistor is cut off. In this case, the cutoff voltage is designated V_{CES}, which stands for collector-emitter voltage with the base *shorted* to the emitter.

On the other hand, if a current source drives the base, the base appears open when the transistor is cut off. The cutoff voltage in this case is symbolized by V_{CEO}, the collector-emitter voltage with the base *open*.

Data sheets often list both values. For instance, the data sheet of a D42C (a 12.5-W power transistor) lists a V_{CES} of 40 V and a V_{CEO} of 30 V. Therefore, the transistor can withstand a maximum collector-emitter voltage of 40 V when the base is driven by a voltage source. If a current source is involved, the collector diode can only withstand 30 V. In case you're not sure whether a voltage or current source is involved, use the lower of the two breakdown voltages as the maximum rating.

EXAMPLE 9-1

Figure 9-3a is the outline of a D42C. The metal tab can be fastened to the chassis to increase the maximum power rating. The data sheet of a D42C gives the following maximum ratings: $I_C = 5$ A, $V_{CEO} = 30$ V, and $V_{CES} = 40$ V. Show that none of these ratings is exceeded in Fig. 9-3b.

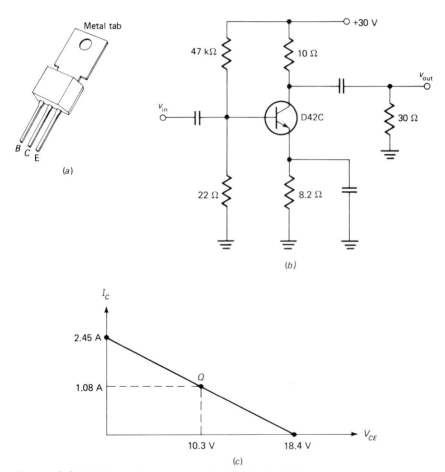

Figure 9-3 (a) Power-tab transistor. (b) Circuit. (c) AC load line.

Transistor Circuit Approximations

SOLUTION

Visualize the dc equivalent circuit. The dc voltage across the 22-Ω biasing resistor is

$$V_2 = \frac{R_2}{R_1 + R_2} \, V_{CC} = \frac{22 \, \Omega}{69 \, \Omega} \, 30 \, V = 9.57 \, V$$

The dc emitter current is

$$I_E = \frac{V_2 - V_{BE}}{R_E} = \frac{9.57 \, V - 0.7 \, V}{8.2 \, \Omega} = 1.08 \, A$$

Since collector current approximately equals emitter current,

$$I_{CQ} = 1.08 \, A$$

and

$$V_{CEQ} = V_{CC} - I_{CQ}(R_C + R_E) = 30 \, V - (1.08 \, A)18.2 \, \Omega = 10.3 \, V$$

The ac collector resistance is

$$r_C = 10 \, \Omega \, \| \, 30 \, \Omega = 7.5 \, \Omega$$

If the input signal is large enough, the operating point swings over the entire ac load line. At the upper end of the ac load line,

$$I_{C(\text{sat})} = I_{CQ} + \frac{V_{CEQ}}{r_C} = 1.08 \, A + \frac{10.3 \, V}{7.5 \, \Omega} = 2.45 \, A$$

At the lower end of the ac load line,

$$V_{CE(\text{cutoff})} = V_{CEQ} + I_{CQ}r_C = 10.3 \, V + (1.08 \, A \times 7.5 \, \Omega) = 18.4 \, V$$

Figure 9-3c shows the ac load line. Since the D42C has maximum I_C rating of 5 A, there is no danger of exceeding this rating, because at most we can have 2.45 A of collector current. Also, the largest collector voltage occurs at cutoff and equals 18.4 V, which is less than the maximum V_{CEO} rating of 30 V.

156

Incidentally, the dc saturation current is 1.65 A and the dc cutoff voltage is 30 V. Therefore, the dc load line is different from the ac load line. This is to be expected, because the dc equivalent circuit is different from the ac equivalent circuit.

9-2 THE AC LOAD LINE OF AN EMITTER FOLLOWER

Figure 9-4 shows the ac load line of an emitter follower. As you see, the Q point has coordinates of I_{CQ} and V_{CEQ}. The saturation value of collector current is

$$I_{C(\text{sat})} = I_{CQ} + \frac{V_{CEQ}}{r_E} \qquad (9\text{-}5)$$

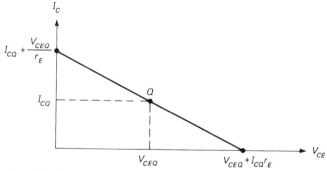

Figure 9-4 AC load of emitter follower.

The cutoff value of collector-emitter voltage is

$$V_{CE(\text{cutoff})} = V_{CEQ} + I_{CQ}r_E \qquad (9\text{-}6)$$

As before, the saturation current must be less than the maximum I_C rating of the transistor. Likewise, the cutoff voltage must be less than the maximum V_{CEO} or V_{CES}, whichever is appropriate. If uncertain, use the lower of the two values.

EXAMPLE 9-2

The transistor of Fig. 9-5a has an I_C rating of 1 A and a V_{CEO} of 20 V. Show that neither of these ratings is exceeded during the ac cycle.

Transistor Circuit Approximations

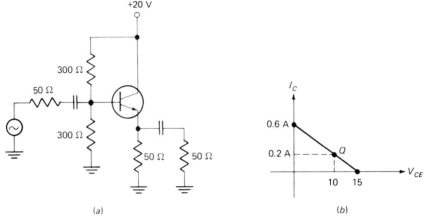

Figure 9-5

SOLUTION
First, get I_{CQ} and V_{CEQ}. Ignoring the V_{BE} drop,

$$I_{CQ} = \frac{10 \text{ V}}{50 \text{ }\Omega} = 0.2 \text{ A}$$

$$V_{CEQ} = 20 \text{ V} - (0.2 \text{ A} \times 50 \text{ }\Omega) = 10 \text{ V}$$

Figure 9-5b shows this Q point.
   In the ac equivalent circuit,

$$r_E = 50 \text{ }\Omega \parallel 50 \text{ }\Omega = 25 \text{ }\Omega$$

The ac saturation current is

$$I_{C(sat)} = I_{CQ} + \frac{V_{CEQ}}{r_C} = 0.2 \text{ A} + \frac{10 \text{ V}}{25 \text{ }\Omega} = 0.6 \text{ A}$$

The ac cutoff voltage is

$$V_{CE(cutoff)} = V_{CEQ} + I_{CQ}r_C = 10 \text{ V} + (0.2 \text{ A} \times 25 \text{ }\Omega) = 15 \text{ V}$$

Figure 9-5b shows the ac load line with the saturation and cutoff values. At no time during the ac cycle are the maximum I_C rating (1 A) and the maximum V_{CEO} rating (20 V) exceeded.

158

9-3 THE OPTIMUM Q POINT

If you overdrive an amplifier, the output signal will be clipped on either or both peaks. Figure 9-6a shows clipping on the negative half cycle of ac source voltage. Since the Q point is closer to cutoff than to saturation, the instantaneous operating point hits the cutoff point before the saturation point. Because of this, we get *cutoff clipping* as shown.

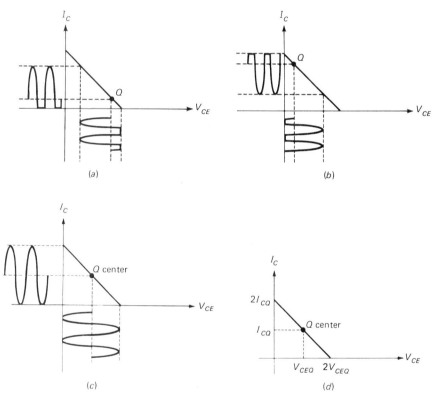

Figure 9-6 *(a)* Cutoff clipping. *(b)* Saturation clipping. *(c)* Maximum signal swing. *(d)* Centered Q point.

If the Q point is too high, that is, closer to saturation than to cutoff, we get *saturation clipping* as shown in Fig. 9-6b. On the positive half cycle of ac source voltage, the instantaneous operating point drives into the saturation point and results in positive clipping of collector current.

To get the maximum unclipped signal, we must locate the Q point at the center of the ac load line (Fig. 9-6c). In this way, the instantaneous operating point can swing equally in both directions before clipping occurs.

Transistor Circuit Approximations

With the right size of input signal, we can get the maximum possible unclipped output signal.

Definition of class A

All along, we have said a transistor should stay in the active region throughout the ac cycle. To distinguish this operation from other kinds, we call it *class A* operation. In terms of the ac load line, class A operation means that no clipping occurs at either end of the ac load line. If we did get clipping, the operation would no longer be called class A operation.

In a class A amplifier the best place to locate the Q point is in the center of the ac load line as shown in Fig. 9-6c. The reason is clear. When the Q point is in the middle of the ac load line, we can get the largest possible unclipped signal.

Centered Q point for CE amplifier

Figure 9-6d shows a centered Q point. Simple geometry tells us the saturation current must be twice as large as I_{CQ} and cutoff voltage twice as large as V_{CEQ}; otherwise, the Q point would not be centered. In symbols,

$$I_{C(\text{sat})} = 2I_{CQ}$$

and

$$V_{CE(\text{cutoff})} = 2V_{CEQ}$$

In a CE amplifier, the saturation current is

$$I_{C(\text{sat})} = I_{CQ} + \frac{V_{CEQ}}{r_c}$$

When the Q point is centered, this becomes

$$2I_{CQ} = I_{CQ} + \frac{V_{CEQ}}{r_c}$$

Rearranging this equation gives

$$r_c = \frac{V_{CEQ}}{I_{CQ}} \qquad \text{(centered } Q\text{)} \qquad \textbf{(9-7)}$$

This final result is important because it allows you to test whether or not the Q point is centered. In other words, the Q point is centered in a CE amplifier only when the ac collector resistance equals the ratio of V_{CEQ} to I_{CQ}. You will find this equation useful in analyzing power amplifiers.

Centered Q point for emitter follower

By a similar argument, an emitter follower has a centered Q point only when

$$r_E = \frac{V_{CEQ}}{I_{CQ}} \qquad \text{(centered } Q\text{)} \qquad \textbf{(9-8)}$$

When this condition is satisfied, an emitter follower produces the maximum possible unclipped output signal.

EXAMPLE 9-3

In Example 9-1, we had $r_C = 7.5$ Ω, $V_{CEQ} = 10.3$ V, and $I_{CQ} = 1.08$ A. Was the Q point centered in the CE amplifier?

SOLUTION

$$r_C = 7.5 \text{ Ω}$$

and

$$\frac{V_{CEQ}}{I_{CQ}} = \frac{10.3 \text{ V}}{1.08 \text{ A}} = 9.54 \text{ Ω}$$

Since r_C does not equal V_{CEQ}/I_{CQ}, the Q point was not centered.

9-4 POWER FORMULAS

What is the maximum power you can get out of a class A amplifier? How much power must a transistor dissipate? These are some of the questions answered in this section.

Quiescent power dissipation

Figure 9-7a shows the Q point for a CE amplifier. With no ac input signal, the transistor has a power dissipation of

$$P_{DQ} = V_{CEQ}I_{CQ} \qquad \textbf{(9-9)}$$

For instance, the D42C of Example 9-1 has $V_{CEQ} = 10.3$ V and $I_{CQ} = 1.08$ A. With no ac input signal, the power dissipation of the transistor is

$$P_{DQ} = 10.3 \text{ V} \times 1.08 \text{ A} = 11.1 \text{ W}$$

161

Transistor Circuit Approximations

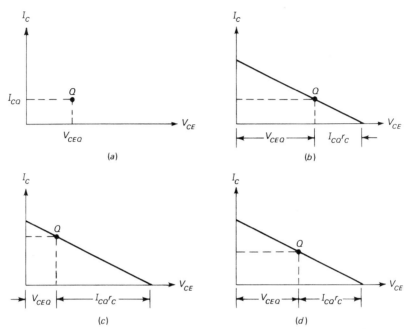

(a)

(b)

(c)

(d)

Figure 9-7 *(a)* Any Q point. *(b)* Q point below center. *(c)* Q point above center. *(d)* Q point at center.

To avoid damage, P_{DQ} must be less than the maximum power rating of the transistor. A D42C can handle up to 12.5 W; therefore, 11.1 W of quiescent power dissipation is all right.

Q point off-center

Figure 9-7*b* shows the ac load line with the Q point closer to cutoff than to saturation. This means cutoff clipping occurs first. The largest unclipped signal in this case has a peak of

$$V_P = I_{CQ}r_c$$

In Fig. 9-7*c*, the Q point is closer to saturation. This means the largest unclipped output signal has a peak of

$$V_P = V_{CEQ}$$

When the Q point is not centered, the maximum output power for an unclipped signal is

162

$$P_0 = \frac{(V_{RMS})^2}{r_C} = \frac{(V_P/\sqrt{2})^2}{r_C}$$

or $$P_0 = \frac{V_P^2}{2r_C} \qquad \textbf{(9-10)}$$

where $V_P = I_{CQ}r_C$ or V_{CEQ}, whichever is smaller.

As an example, look at Fig. 9-3c. The Q point is closer to cutoff; therefore, the largest unclipped output signal has a peak of

$$V_P = 18.4 \text{ V} - 10.3 \text{ V} = 8.1 \text{ V}$$

The maximum output power is

$$P_0 = \frac{V_P^2}{2r_C} = \frac{(8.1 \text{ V})^2}{2(7.5 \text{ }\Omega)} = 4.37 \text{ W} \qquad \textbf{(9-11)}$$

Centered Q point

Figure 9-7d shows a centered Q point. Since the peak swing is equal above and below the Q point,

$$V_P = V_{CEQ} = I_{CQ}r_C$$

Substituting into Eq. (9-10) gives

$$P_0 = \frac{V_P^2}{2r_C} = \frac{V_{CEQ}I_{CQ}r_C}{2r_C} = \frac{V_{CEQ}I_{CQ}}{2}$$

or $$P_0 = \frac{P_{DQ}}{2} \qquad \textbf{(9-12)}$$

This final result is important. It says the maximum ac output power delivered to r_C always equals half the quiescent power dissipation of the transistor. The Q point must be centered on the ac load line, which is the way most class A power amplifiers are designed.

As an example, if a transistor dissipates 3 W under no-signal conditions, the maximum ac output power it can deliver is 1.5 W. Conversely, if you are trying to build a class A amplifier that delivers 30 W of ac output power, you will need a transistor that can dissipate at least 60 W at the Q point.

163

Emitter follower

The power formulas for an emitter follower are similar to those just given for a CE amplifier. As before,

$$P_{DQ} = V_{CEQ}I_{CQ}$$

For the Q point off the center of the ac load line, maximum voltage swing without clipping is

$$V_P = V_{CEQ} \text{ or } I_{CQ}r_E$$

whichever is smaller. The maximum ac output power is

$$P_0 = \frac{V_P^2}{2r_E}$$

When the Q point is centered, the maximum ac output power is half the quiescent power dissipation of the transistor:

$$P_0 = \frac{P_{DQ}}{2}$$

EXAMPLE 9-4

Figure 9-5b is the ac load line of the emitter follower analyzed earlier. What is the quiescent power dissipation of the transistor? The maximum ac output power delivered to r_E?

SOLUTION

The quiescent power dissipation is

$$P_{DQ} = V_{CEQ}I_{CQ} = 10 \text{ V} \times 0.2 \text{ A} = 2 \text{ W}$$

The Q point is not centered. The peak swing without clipping is $V_P = 5$ V. So, the maximum ac output power is

$$P_0 = \frac{V_P^2}{2r_C} = \frac{(5 \text{ V})^2}{2(25 \text{ }\Omega)} = 0.5 \text{ W}$$

9-5 LARGE-SIGNAL GAIN AND IMPEDANCE

When analyzing small-signal amplifiers, we can use r'_e because the signal swing covers only a small part of the ac load line. But in power amplifiers,

164

the signal swing uses most if not all of the ac load line. In this case, we have to use the *large-signal* ac emitter resistance, symbolized by R'_e.

How to get R'_e

Because R'_e is a large-signal characteristic of a power transistor, we need information from the data sheet to get its value; no simple formula like 25 mV/I_E exists for R'_e. Instead, we have to use the *transconductance curve* shown on the data sheet of a power transistor.

The transconductance curve is a graph of I_C versus V_{BE}, similar to Fig. 9-8. R'_e is defined as the ratio of a large change in V_{BE} to a large change in I_E. Since I_C approximately equals I_E,

$$R'_e = \frac{\Delta V_{BE}}{\Delta I_C} \qquad\qquad \textbf{(9-13)}$$

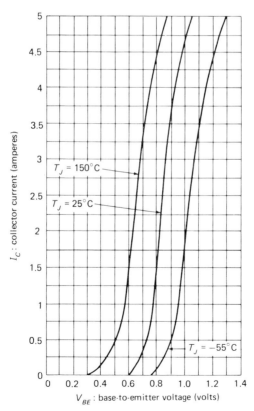

Figure 9-8 Transconductance curves.

Transistor Circuit Approximations

By using the transconductance curve, we can calculate R_e'. For instance, suppose the collector current swings from 0 to 2 A over the ac cycle. For a junction temperature of 25°C (the middle curve), the corresponding swing in base-emitter voltage is from approximately 0.6 to 0.82 V. The value of R_e' is

$$R_e' = \frac{\Delta V_{BE}}{\Delta I_C} = \frac{0.82 \text{ V} - 0.6 \text{ V}}{2 \text{ A} - 0} = 0.11 \ \Omega$$

Large-signal formulas

The derivations of large-signal formulas are similar to those given earlier for small-signal amplifiers. The difference is that R_e' is used instead of r_e', and β_{dc} instead of β. In other words, when we want a large-signal formula for gain or input impedance, all we do is replace r_e' by R_e' and β by β_{dc}.

For instance, the voltage gain of a CE amplifier driven by a small signal is

$$A = \frac{r_C}{r_e'}$$

When this amplifier is driven by a large signal, the voltage gain is

$$A = \frac{r_C}{R_e'}$$

The input impedance looking into the base with a small signal is

$$z_{\text{in(base)}} = \beta r_e'$$

But with large signals, it becomes

$$z_{\text{in(base)}} = \beta_{\text{dc}} R_e'$$

Nonlinear distortion

Transconductance curves like Fig. 9-8 are *nonlinear*. Because of this, a large sinusoidal base voltage in a CE amplifier produces a *nonsinusoidal* collector current as in Fig. 9-9. When this nonsinusoidal current flows through the ac collector resistance, we get a nonsinusoidal output voltage. In other words, when a perfect sine wave goes into a large-signal CE amplifier, an imperfect sine wave comes out. The change in the shape of the signal is called *nonlinear distortion*.

166

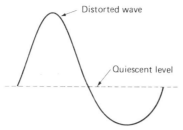

Figure 9-9 Nonlinear distortion.

The nonlinearity of the emitter diode is the cause of the distortion.   One way to reduce its effect is with a swamping resistor.   As you recall, we can insert a resistor between the emitter and ac ground.   In the small-signal case, the voltage gain becomes

$$A = \frac{r_C}{r_E + r_e'}$$

For large signals

$$A = \frac{r_C}{r_E + R_e'}$$

As long as r_E is much greater than R_e', the emitter diode is swamped out and the nonlinear distortion is greatly reduced.

Emitter follower

The emitter follower is a popular circuit when it comes to large-signal opera- tion, because of its low nonlinear distortion.   The large-signal voltage gain of an emitter follower is

$$A = \frac{r_E}{r_E + R_e'}$$

To keep the voltage gain near unity, a designer makes r_E much greater than R_e'.   This results in heavy swamping of the emitter diode and means that nonlinear distortion is very small.

EXAMPLE 9-5

An emitter follower has $r_E = 10 \ \Omega$ and $R_e' = 0.11 \ \Omega$.   The transistor has the h_{FE} curve shown in Fig. 9-10.   If the collector current swings from 0 to 2 A, what do A and $z_{\text{in(base)}}$ equal?

167

Transistor Circuit Approximations

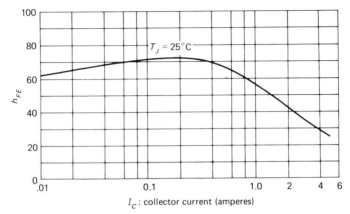

Figure 9-10 Variation of h_{FE} with collector current.

SOLUTION

The voltage gain is

$$A = \frac{r_E}{r_E + R'_e} = \frac{10 \ \Omega}{10 \ \Omega + 0.11 \ \Omega} = 0.989$$

Because large-signal analysis is based on peak-to-peak swings, the correct value of β_{dc} to use is the value for the maximum collector current—in other words, the value of β_{dc} for I_C equal to 2 A. In Fig. 9-10, h_{FE} is approximately 42 when I_C is 2 A. The ac impedance looking into the base is

$$z_{in(base)} = \beta_{dc}(r_E + R'_e) = 42(10.11 \ \Omega) = 425 \ \Omega$$

9-6 TRANSISTOR POWER RATINGS

The temperature at the collector junction places a limit on the allowable power dissipation of a transistor. Depending on the transistor type, a junction temperature from 150 to 200°C will destroy the transistor. Data sheets specify this maximum junction temperature as $T_{J(max)}$.

Ambient temperature

The heat produced at the junction passes through the transistor case—either a metal or plastic housing—and radiates to the surrounding air. The temperature of this air, known as the *ambient* temperature, is around 25°C, but it can be much higher on hot days. Also, the ambient temperature may be much higher inside a piece of electronics equipment.

168

Thermal resistance

The *thermal resistance* θ_{JA} is a physical property of the transistor and its case; θ_{JA} is the resistance to the flow of heat from the junction to the surrounding air. A low thermal resistance means it's easy for heat to flow from the junction to the surrounding air. In general, the larger the transistor case, the lower the thermal resistance. For example, the thermal resistance of a 2N3904 is

$$\theta_{JA} = 357°C/W$$

The 2N3719 has a larger case and a lower thermal resistance of

$$\theta_{JA} = 175°C/W$$

This implies that a 2N3719 radiates heat more easily than a 2N3904.

Junction temperature

Courses in thermodynamics prove the following equation for junction temperature:

$$T_J = T_A + \theta_{JA}P_D \qquad\qquad \textbf{(9-14)}$$

where T_J = junction temperature
$\quad T_A$ = ambient temperature
$\quad \theta_{JA}$ = thermal resistance
$\quad P_D$ = power dissipation
As an example, here's how to calculate the junction temperature of a 2N3904 with an ambient temperature of 25°C and a power dissipation of 0.2 W. Using the θ_{JA} given earlier,

$$T_J = T_A + \theta_{JA}P_D = 25°C + (357°C/W)(0.2\ W) = 96.4°C$$

As another example, the junction temperature of a 2N3719 with the same ambient temperature and power dissipation is

$$T_J = T_A + \theta_{JA}P_D = 25°C + (175°C/W)(0.2\ W) = 60°C$$

These two examples demonstrate the effect of thermal resistance. Because it has a lower thermal resistance, the 2N3719 runs cooler than the 2N3904 for the same ambient temperature and power dissipation.

169

Derating factor

Data sheets specify the maximum power rating at 25°C. For instance, the 2N3904 has a $P_{D\,(max)}$ of 350 mW for a T_A of 25°C. This means the 2N3904 can have a quiescent power dissipation as high as 350 mW, provided the ambient temperature is 25°C.

If the ambient temperature is greater than 25°C, you have to *derate* (reduce) the power rating. Some data sheets specify a *derating factor* for temperatures above 25°C. For instance, the 2N3904 has a derating factor of 2.8 mW/°C. This means the power rating decreases 2.8 mW for each degree above 25°C. As a formula,

$$P_{D\,(max)} = P_{25} - D(T_A - 25°C) \qquad \text{(9-15)}$$

where P_{25} = power rating at 25°C (ambient)

D = derating factor

Here is an example of how to derate a transistor. The 2N3904 has a maximum power rating of 350 mW for an ambient temperature of 25°C. As mentioned earlier, its derating factor is 2.8 mW/°C. If the ambient temperature is 100°C, the allowable power dissipation decreases to

$$P_{D\,(max)} = 350 \text{ mW} - (2.8 \text{ mW/°C})(100°C - 25°C) = 140 \text{ mW}$$

Sometimes, rather than specify the derating factor, the data sheet shows a graph of $P_{D\,(max)}$ versus temperature. For instance, Fig. 9-11 is the derating curve for a 2N1936. As you can see, the maximum power rating is 4 W from 0 to 25°C. Above 25°C the maximum power rating decreases linearly. If you wanted to use this transistor over a temperature range of 0 to 100°C, the maximum allowable power dissipation would be 2 W.

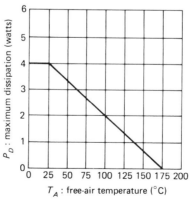

Figure 9-11 Derating curve.

Heat sinks

The lower the thermal resistance, the greater the allowable power dissipation. We can reduce θ_{JA} by attaching a *heat sink* (a mass of metal) to the transistor case. For instance, Fig. 9-12a shows one type of heat sink. The increased surface area of the fins allows heat to escape more easily; this increases the maximum power rating of the transistor.

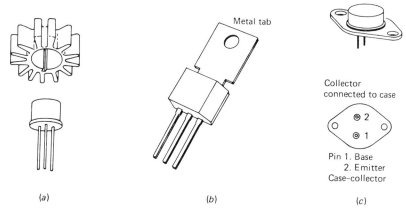

(a) (b) (c)

Figure 9-12 *(a)* Attachable heat sink. *(b)* Power-tab transistor. *(c)* Power transistor.

Figure 9-12b is the outline of a power-tab transistor. The metal tab is fastened to the chassis, which acts like a massive heat sink. Because of this, heat escapes rapidly from the transistor, reducing the junction temperature. This means the transistor can dissipate much more heat before the junction temperature reaches destructive levels (150 to 200°C).

The larger power transistors have the collector connected directly to the case to minimize the thermal resistance between the junction and the case (Fig. 9-12c). The case can then be placed in thermal contact with the chassis for heat-sinking. To prevent the collector from being shorted to ground, a thin mica washer is used between the case and the chassis.

Case-temperature formulas

Data sheets for larger power transistors usually specify the power rating for case temperatures rather than for ambient temperatures, the assumption being that the chassis will be used as a heat sink, rather than the surrounding air. Because of this, Eq. (9-14) can be rewritten as

$$T_J = T_C + \theta_{JC}P_D \qquad \text{(9-16)}$$

Transistor Circuit Approximations

where T_C = case temperature

θ_{JC} = thermal resistance between junction and case

Because θ_{JC} is much smaller than θ_{JA}, the junction temperature is much lower for a transistor using the chassis as a heat sink.

The thermal resistance between the case and the surrounding air is designated θ_{CA}. When θ_{CA} is known, you can calculate the case temperature of a transistor with

$$T_C = T_A + \theta_{CA}P_D \tag{9-17}$$

Using the chassis as a heat sink dramatically lowers the value of θ_{CA}; in turn, this means the case temperature is much lower than it would be without heat-sinking.

The derating formula in terms of case temperature is

$$P_{D(\text{max})} = P_{25} - D(T_C - 25°C) \tag{9-18}$$

where $P_{D(\text{max})}$ = power rating at 25°C (case)

D = derating factor

T_C = case temperature

The foregoing formula is usually plotted on the data sheet of larger power transistors. For instance, Fig. 9-13 shows the derating graph for a 2N5877. The power rating is 150 W from 0 to 25°C; then, it decreases linearly until it reaches zero at 200°C.

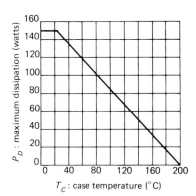

Figure 9-13

EXAMPLE 9-6

A 2N6326 has a derating factor of 1.14 W/°C above case temperatures of 25°C. The maximum power rating at 25°C (case temperature) is 200 W. What is the maximum power rating for a case temperature of 100°C?

172

SOLUTION

Substitute the given values into Eq. (9-18):

$$P_{D(\text{max})} = 200 \text{ W} - (1.14 \text{ W}/^\circ\text{C})(100^\circ\text{C} - 25^\circ\text{C}) = 115 \text{ W}$$

GLOSSARY

ac load line A linear graph of collector current versus collector-emitter voltage for a transistor driven by an ac signal.

class A Operation of the transistor in the active region. Equivalent to operating a transistor anywhere between ac saturation and ac cutoff.

cutoff clipping Clipping of the output signal when a large ac signal drives the transistor into cutoff.

derating factor The decrease in maximum power rating of a transistor for each degree rise in ambient or junction temperature.

heat sink A mass of metal in thermal contact with the transistor case. The heat sink absorbs transistor heat more rapidly than the surrounding air.

large-signal operation There is no hard-and-fast definition of large-signal operation, but it usually means swinging over most of the ac load line.

nonlinear distortion A perfect amplifier enlarges a signal without changing its shape. Any deviation in shape caused by the nonlinearity of the emitter diode is called nonlinear distortion.

power transistor One intended for large-signal operation. Power transistors are arbitrarily defined as those with a maximum power rating of 1 W or more.

saturation clipping Clipping of the output signal when a large ac input signal drives the transistor into saturation.

small-signal transistor Arbitrarily, these are transistors with a maximum power rating of less than 1 W. A small-signal transistor operates over a small part of the ac load line. Typically, such a transistor is used near the front end of a system where the input signal is weak.

thermal resistance The resistance to the flow of heat between two points. With transistors, the most important thermal resistances are θ_{JA}, θ_{JC}, and θ_{CA}.

transconductance curve A graph of collector current versus base-emitter

173

voltage. Data sheets for power transistors often include a transconductance curve.

SELF-TESTING REVIEW

Read each of the following and provide the missing words. Answers appear at the beginning of the next question.

1. The quiescent point is the operating point when no _____ signal drives the amplifier. To find the Q point, you need to analyze the _____ equivalent circuit. To locate the saturation and cutoff points on the ac load line, you must analyze the _____ equivalent circuit.

2. *(input, dc, ac)* V_{CES} is the collector-emitter voltage with the base _____ to the emitter. V_{CEO} is the collector-emitter voltage with the base _____. Data sheets of power transistors often list the maximum V_{CES} and V_{CEO}. When in doubt about which to use, take the _____ of the two values.

3. *(shorted, open, lower)* When the Q point is below the center of the ac load line, _____ clipping occurs first. If the Q point is above the center, _____ clipping occurs first. To get the maximum un-clipped output signal, the Q point should be at the _____ of the ac load line.

4. *(cutoff, saturation, center)* Class A operation means operating the transistor in the _____ region. This implies not driving the transistor into _____ or _____. The best place to locate the Q point of a class A amplifier is in the _____ of the ac load line.

5. *(active, saturation, cutoff, center)* To find out if the Q point of a CE amplifier is centered, check the value of r_C compared to V_{CEQ}/I_{CQ}. These quantities are _____ when the Q point is at the center of the ac load line. With an emitter follower, check the value of _____ compared to V_{CEQ}/I_{CQ}.

6. *(equal, r_E)* The quiescent power dissipation of a transistor equals the product of _____ and _____. For a centered Q point, the maximum ac output power equals half of the _____ power dissipation.

7. *(V_{CEQ}, I_{CQ}, quiescent)* To get the value of R'_e, we need to use the _____ curve, usually found on the data sheets of power transistors. To convert small-signal formulas to large-signal formulas, replace r'_e by _____ and β by _____.

8. *(transconductance, R'_e, β_{dc})* Transconductance curves are nonlinear. Because of this, a sinusoidal base-emitter voltage produces a _____

174

collector current. The change in the shape of the signal is called _____ distortion.

9. *(nonsinusoidal, nonlinear)* You can reduce the nonlinear distortion of a CE amplifier by using a _____ resistor between the emitter and ac ground. The typical emitter follower has heavy swamping, which means very small _____ distortion.

10. *(swamping, nonlinear)* θ_{JA} is the resistance to the flow of _____ from the junction to the surrounding air. A low thermal resistance means it's _____ for heat to flow from the junction to the surrounding air.

11. *(heat, easy)* If the ambient or case temperature is above 25°C, you have to _____ the maximum power rating of the transistor. Some data sheets specify the _____ factor; others give a graph of maximum power rating versus _____.

12. *(derate, derating, temperature)* A heat sink is a mass of metal that absorbs heat from the transistor. A power-tab transistor has a metal tab you can fasten to the chassis, which acts like a massive heat sink. The larger power transistors have the collector connected to the case; a thin mica washer prevents a collector-to-case short, but places the case in close thermal contact with the chassis.

PROBLEMS

9-1. In Fig. 9-14*a*, calculate the quiescent collector current and the collector-

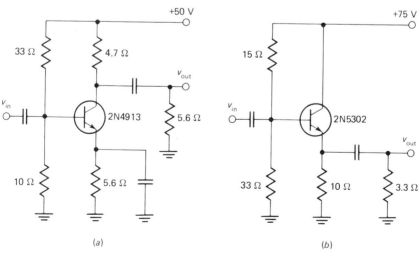

(a)   (b)

Figure 9-14

emitter voltage. Also work out the ac saturation current and the ac cutoff voltage.

9-2. The 2N4913 of Fig. 9-14a has these absolute maximum ratings: $V_{CEO} = 40$ V and $I_C = 15$ A. Show that neither of these ratings is exceeded when the entire ac load line is used.

9-3. The 2N5302 of Fig. 9-14b has a maximum collector-emitter voltage rating of 60 V. If cutoff clipping occurs, what does the collector-emitter voltage equal? Is the maximum rating exceeded?

9-4. In Fig. 9-14b, the 2N5302 has a maximum collector current rating of 50 A. What does the ac saturation current equal? Is this all right as far as the maximum current rating is concerned?

9-5. A CE amplifier has $r_C = 5$ Ω, $V_{CEQ} = 20$ V, and $I_{CQ} = 4$ A. Is the Q point centered?

9-6. Test each of the following CE amplifier values for a centered Q point:

a. $V_{CEQ} = 25$ V, $I_{CQ} = 5$ V, $r_C = 4$ Ω
b. $V_{CEQ} = 5$ V, $I_{CQ} = 2$ A, $r_C = 2.5$ Ω
c. $r_C = 12$ Ω, $V_{CEQ} = 30$ V, $I_{CQ} = 2.5$ A
d. $I_{CQ} = 3$ A, $V_{CEQ} = 24$ V, $r_C = 6$ Ω

9-7. What is the quiescent power dissipation for the transistor of Fig. 9-14a? What is the maximum unclipped ac output power?

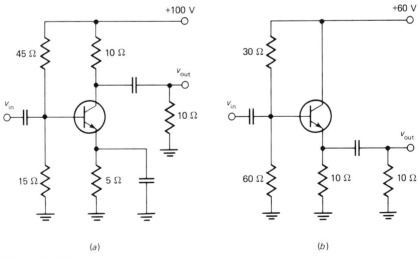

(a)   (b)

Figure 9-15

9-8. In Fig. 9-14*b*, what is the quiescent power dissipation of the transistor? The maximum unclipped ac output power?

9-9. The Q point of Fig. 9-15*a* is approximately centered on the ac load line. What is the quiescent power dissipation of the transistor? The maximum ac output power?

9-10. You want to build a CE amplifier that delivers 40 W of ac output power. What is the maximum power rating needed for the transistor?

9-11. The Q point is approximately at the center of the ac load line in Fig. 9-15*b*. What is the maximum ac output power? The ac power in the final 10-Ω resistor?

9-12. A transistor has the transconductance curves of Fig. 9-8. Calculate its R'_e for each of these conditions:

a. $T_J = 25°C$ and $I_C = 0$ to 4 A
b. $T_J = 150°C$ and $I_C = 0.5$ to 5 A
c. $T_J = -55°C$ and $I_C = 0.25$ to 4.5 A

9-13. If the transistor of Fig. 9-15*a* has $R'_e = 0.07$ Ω and $\beta_{dc} = 56$, what is the large-signal voltage gain? The input impedance of the stage?

9-14. $R'_e = 0.1$ Ω and $\beta_{dc} = 48$ in Fig. 9-15*b*. What is the voltage gain? The input impedance of the stage?

9-15. A D44C power-tab transistor has $\theta_{JA} = 75°C/W$. If this transistor dissipates 1 W for an ambient temperature of 25°C, what is the junction temperature?

9-16. Solve the foregoing problem for $T_A = 50°C$.

9-17. The 2N4401 has a maximum power rating of 310 mW for an ambient temperature of 25°C. The derating factor is 2.81 mW/°C. What is the maximum power rating for an ambient temperature of 100°C?

9-18. The θ_{JC} of a D44C is 4.2°C/W. If the transistor dissipates 20 W when the case temperature is 25°C, what is the junction temperature?

9-19. A power transistor uses the chassis for a heat sink. If $\theta_{CA} = 1°C/W$, what is the case temperature for $T_A = 25°C$ and $P_D = 100$ W?

9-20. A TIP29 has a power rating of 30 W for a case temperature of 25°C. If the derating factor is 0.24 W/°C, what is the power rating for a case temperature of 70°C?

Chapter 10

Other Power Amplifiers

The class B push-pull amplifier is a two-transistor circuit that can deliver more ac output power than a class A amplifier. Another advantage of class B operation is that no-signal current drain from the power supply is very small. This is important for battery-operated equipment.

The class C amplifier can deliver more ac output power than class A or class B. But the class C amplifier has to be tuned to a resonant frequency. Because of this, class C amplifiers are inherently narrow-band circuits that work at radio frequencies (above 20 kHz).

Besides class A, B, and C operation, the transistor can be used as a switch. This means using only two points on the ac load line, typically cutoff and saturation. One important industrial application of switching operation is controlling large amounts of current through heavy loads.

10-1 CLASSES OF OPERATION

The transistor of a class A amplifier operates in the active region throughout the ac cycle. This means collector current flows for 360° as shown in Fig. 10-1a. At no point in the cycle is the transistor driven into saturation or cutoff.

Class B

In a class B circuit, the transistor operates in the active region for only half the cycle. During the other half cycle, the transistor is cut off. This means collector current exists for 180° as shown in Fig. 10-1b. Normally, a second transistor supplies the missing half cycle. In other words, two transistors

178

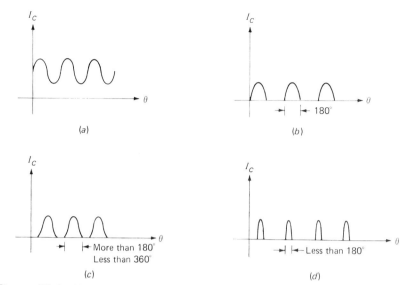

Figure 10-1 Classes of operation. *(a)* A. *(b)* B. *(c)* AB. *(d)* C.

each operating class B form a *push-pull* amplifier; one transistor takes care of the positive half cycle, and the other handles the negative half cycle.

Class AB

Class AB is between class A and class B. The transistor of a class AB circuit is in the active region for more than half a cycle but less than a whole cycle. For class AB operation, collector current exists for more than 180° but less than 360° as shown in Fig. 10-1*c*.

Class C

Class C operation means collector current flows for less than 180° as shown in Fig. 10-1*d*. In a practical class C circuit, the current exists for much less than 180° and looks like narrow pulses. As will be discussed later, when narrow current pulses like these drive a high-*Q* resonant circuit, the voltage across the tank is almost a perfect sine wave.

10-2 THE CURRENT MIRROR

Figure 10-2*a* shows a *current mirror,* extensively used in linear integrated circuits. The current directions are for conventional current. If you prefer

179

Transistor Circuit Approximations

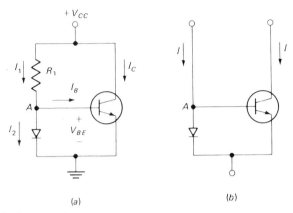

Figure 10-2 Current mirror.

electron flow, visualize the currents in the opposite direction. Current I_1 flows down through R_1 and splits into I_2 and I_B. Current I_2 flows through a *compensating* diode, and I_B flows into the base.

If the current-voltage curve of the compensating diode is identical to the transconductance curve of the transistor, the diode current equals the collector current. In symbols,

$$I_2 = I_C$$

At node A,

$$I_1 = I_2 + I_B$$

or

$$I_1 = I_C + I_B$$

Since base current is much smaller than collector current, this reduces to

$$I_1 \cong I_C$$

This is important. It says the collector current equals the current entering node A.

Figure 10-2b emphasizes this point. Current I into node A sets up an equal collector current I. Think of the circuit as a mirror; the current into node A is reflected into the collector circuit, where an equal current appears. This is why the diode-transistor circuit of Fig. 10-2b is known as a current mirror.

180

EXAMPLE 10-1

What is the collector current in Fig. 10-3a?

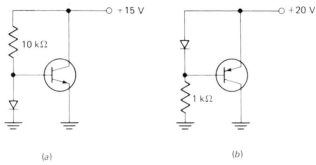

(a)

(b)

Figure 10-3 *(a) npn* mirror. *(b) pnp* mirror.

SOLUTION

The current through the resistor is

$$I = \frac{V_{CC} - V_{BE}}{R} = \frac{15 \text{ V} - 0.7 \text{ V}}{10 \text{ k}\Omega} = 1.43 \text{ mA}$$

Therefore, the collector current is

$$I_C = 1.43 \text{ mA}$$

EXAMPLE 10-2

What is the collector current in Fig. 10-3b?

SOLUTION

A *pnp* mirror works the same as an *npn* mirror. The current through the resistor is

$$I = \frac{20 \text{ V} - 0.7 \text{ V}}{1 \text{ k}\Omega} = 19.3 \text{ mA}$$

So, the collector current is

$$I_C = 19.3 \text{ mA}$$

10-3 BIASING A CLASS B PUSH-PULL AMPLIFIER

The most difficult thing about a class B push-pull amplifier is setting up a stable Q point. It is much more difficult than in a class A circuit, because V_{BE} becomes a crucial quantity.

Voltage-divider bias

Figure 10-4a shows voltage-divider bias for a class B push-pull emitter follower. The transistors need to be *complementary,* meaning similar transconductance curves, maximum ratings, etc. For instance, the 2N3904 and 2N3906 are complementary, the first being an *npn* transistor and the second a *pnp;* both have almost the same maximum ratings, transconductance curves, and so on. Complementary transistors are commercially available for almost any application.

In Fig. 10-4a, collector and emitter currents are in the same direction through both transistors. Because of the series connection, the collector currents are equal and the collector-emitter voltage of each transistor is half the supply voltage.

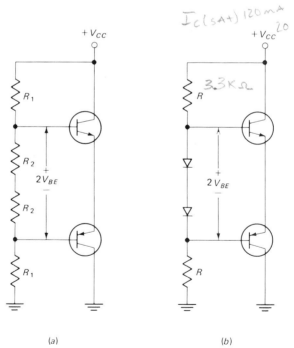

$I_{c(sAt)}\ 120\ mA$

$+V_{cc}\ 20$

$3.3K\Omega$

(a)   (b)

Figure 10-4 Class B push-pull circuit. *(a)* Voltage-divider bias. *(b)* Current-mirror bias.

182

To set up class B operation, we need to bias each emitter diode at the knee, where it just begins to conduct. This means roughly 0.7 V across each R_2. The correct value of V_{BE} may be slightly lower or higher, depending on the transistor type, the temperature, etc. Unfortunately, the transconductance curve is so steep in the knee area that even a 0.1-V error can produce a huge change in collector current. Since published transconductance curves have tolerance, the correct bias voltage for a particular transistor has to be found experimentally, using an adjustable R_1 or R_2.

Voltage-divider bias like Fig. 10-4a has been used, but it's one of the worst ways to bias a class B push-pull amplifier. Any practical circuit has to contend with large changes in junction temperature. As this temperature increases, the barrier potential of the emitter diodes decreases about 2.5 mV per degree. Therefore, the fixed voltage produced by R_2 will overbias the emitter diode and produce large increases in collector current when the junction temperature rises.

Current-mirror bias

Figure 10-4b shows *current-mirror bias,* the main method used to bias class B push-pull amplifiers. The transistors are complementary, and the curves of the compensating diodes match the transconductance curves of the transistors. Therefore, the upper half of the circuit is an *npn* mirror, and the lower half is a *pnp* mirror. Note that the same collector current exists in both transistors. Furthermore, this collector current is a reflection of the current through the biasing resistors.

For current-mirror bias to work properly, the curves of the compensating diodes must match the transconductance curves of the transistors over a large temperature range. In discrete circuits, this is hard to do because of diode and transistor tolerances; so you have to settle for less-than-perfect mirror action. But current-mirror bias is outstanding when it comes to linear integrated circuits. Since the compensating diodes and the emitter diodes are on the same chip, they have almost identical characteristics; this means almost perfect mirror action over a large temperature range.

Crossover distortion

It is important to have a small collector current under no-signal conditions. Otherwise, we run the risk of *crossover distortion.* Here is the idea. Suppose no bias at all is applied to the emitter diodes. Then, the incoming ac signal has to rise to about 0.7 V to overcome the barrier potential (see Fig. 10-5a). Because of this, no collector current exists when the signal is less than 0.7 V. The action on the other half cycle is complementary; the other transistor

183

Transistor Circuit Approximations

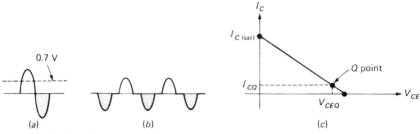

Figure 10-5 *(a)* Overcoming the barrier potential. *(b)* Crossover distortion. *(c)* Slight forward bias.

does not turn on until the ac input signal is more negative than -0.7 V. For this reason, if no bias at all is applied to the emitter diodes, the output of a class B push-pull amplifier looks like Fig. 10-5*b*.

This output signal is distorted; it is no longer a sine wave because of the clipping action between half cycles. Since the clipping occurs between the time one transistor shuts off and the other comes on, we call it crossover distortion.

To eliminate crossover distortion, we need to apply a slight forward bias to each emitter diode. This means locating the Q point slightly above cutoff as shown in Fig. 10-5*c*. As a guide, an I_{CQ} from 1 to 5 percent of $I_{C(sat)}$ is enough to eliminate crossover distortion. Since I_{CQ} is a reflection of the current through the biasing resistors, its value in Fig. 10-4*b* is

$$I_{CQ} = \frac{V_{CC} - 2V_{BE}}{2R} \qquad \textbf{(10-1)}$$

EXAMPLE 10-3

What is the collector current in Fig. 10-4*b* if $V_{CC} = 20$ V and $R = 3.3$ kΩ? If $I_{C(sat)} = 120$ mA, what does $I_{CQ}/I_{C(sat)}$ equal?

SOLUTION

The quiescent collector current is

$$I_{CQ} = \frac{V_{CC} - 2V_{BE}}{2R} = \frac{20 \text{ V} - 1.4 \text{ V}}{6.6 \text{ k}\Omega} = 2.82 \text{ mA}$$

This is small compared to $I_{C(sat)}$ because the ratio is

$$\frac{I_{CQ}}{I_{C(sat)}} = \frac{2.82 \text{ mA}}{120 \text{ mA}} = 0.0235$$

184

which is equivalent to 2.35 percent. As mentioned earlier, to avoid crossover distortion, I_{CQ} ought to be from 1 to 5 percent of $I_{C(sat)}$.

A final point: Strictly speaking, we have class AB operation rather than class B because of the small I_{CQ}. But the operation is so close to class B that most people continue to call the circuit a class B push-pull amplifier.

10-4 CLASS B PUSH-PULL EMITTER FOLLOWER

Figure 10-6 shows a class B push-pull amplifier. Current-mirror bias sets the Q point slightly above cutoff to prevent crossover distortion. The positive half cycle of input voltage is coupled through the capacitor and through the diodes to the bases. This positive-going voltage will turn on the upper transistor and shut off the lower one. The emitter voltage of the upper transistor now follows the base voltage. The output capacitor couples the positive half cycle to the load resistance R_L.

The action is complementary on the negative half cycle. This means the upper transistor shuts off and the lower one comes on. The lower transistor now acts like an emitter follower, and the output capacitor couples the negative half cycle to the load resistance. The final signal across the load is symmetrical because the two amplified half cycles are recombined.

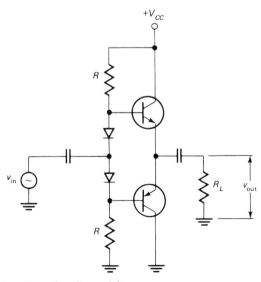

Figure 10-6 Class B push-pull amplifier.

Transistor Circuit Approximations

Impedance and gain formulas

The large-signal input impedance looking into the base of the active transistor is

$$z_{in(base)} = \beta_{dc}(R_L + R'_e)\qquad\text{(10-2)}$$

The large-signal voltage gain is

$$A = \frac{R_L}{R_L + R'_e}\qquad\text{(10-3)}$$

As before, the large-signal power gain equals the product of large-signal voltage gain and current gain:

$$G = \frac{\beta_{dc}R_L}{R_L + R'_e}\qquad\text{(10-4)}$$

Output power

Figure 10-7 shows the ideal ac load line for a class B push-pull amplifier. It is ideal because it neglects $V_{CE(sat)}$ and I_{CQ}. In a real amplifier, the saturation point does not quite touch the vertical axis, and the Q point is slightly above the horizontal axis.

Here is the main idea. Figure 10-7 shows the largest current and voltage waveforms we can get with one transistor of a class B push-pull amplifier; the other transistor produces the dotted half cycles. For this reason, the ac

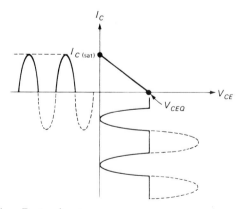

Figure 10-7 Class B signal swing.

186

output voltage has a peak value of V_{CEQ}, and the ac output current has a peak value of $I_{C(sat)}$. Therefore, the maximum ac output power is

$$P_O = V_{RMS}I_{RMS} = \frac{V_{CEQ}}{\sqrt{2}}\frac{I_{C(sat)}}{\sqrt{2}}$$

or
$$P_O = \frac{V_{CEQ}I_{C(sat)}}{2} \qquad \text{(10-5)}$$

Power dissipation

Under no-signal conditions, the transistors of a class B push-pull amplifier are *idling,* because only a small trickle of current passes through them. In this case, the power dissipation in each transistor is negligible. But when a signal is present, large currents are in each transistor, causing significant power dissipation. The worst-case power dissipation is too complicated to derive here, but with calculus it can be shown that

$$P_D \cong \frac{V_{CEQ}I_{C(sat)}}{10} \qquad \text{(10-6)}$$

With Eq. (10-5), this reduces to

$$P_D \cong \frac{P_O}{5} \qquad \text{(10-7)}$$

This means the power rating of each transistor in a class B push-pull amplifier has to be greater than one-fifth of the maximum ac output power. For instance, to deliver 100 W of ac output power, each transistor must have a power rating of at least 20 W.

Darlington and complementary pairs

Darlington pairs can increase the input impedance of a class B push-pull emitter follower (see Fig. 10-8a). Since the Darlington beta is much higher, the input impedance is much higher. But the disadvantage of using Darlingtons is four compensating diodes instead of two, which makes it more difficult to match the compensating diodes to the emitter diodes.

This is where *complementary pairs* come in. In Fig. 10-8b, the upper pair acts like an *npn* transistor, and the lower pair like a *pnp* transistor. The effective beta of each pair equals the product of the individual betas,

187

Transistor Circuit Approximations

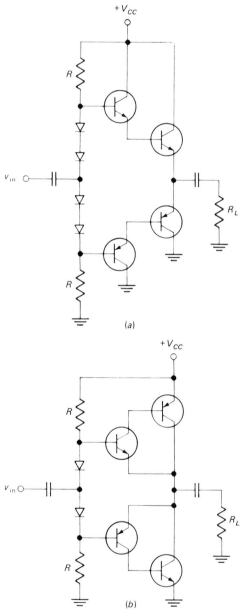

Figure 10-8 *(a)* Darlington pairs. *(b)* Complementary pairs.

the same as a Darlington. But the big advantage of complementary pairs is simpler compensation; two diodes instead of four. This means it's easier to match compensating diodes to emitter diodes.

EXAMPLE 10-4

Calculate the maximum ac output power in Fig. 10-9. Also, work out the required power rating of the transistors.

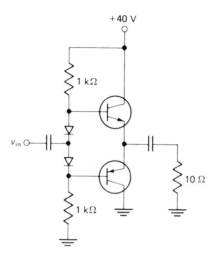

Figure 10-9

SOLUTION

V_{CEQ} equals half the supply voltage:

$$V_{CEQ} = \frac{V_{CC}}{2} = \frac{40 \text{ V}}{2} = 20 \text{ V}$$

The general formula for ac saturation current still applies:

$$I_{C(sat)} = I_{CQ} + \frac{V_{CEQ}}{r_E}$$

In a class B amplifier, I_{CQ} is only 1 to 5 percent of $I_{C(sat)}$, small enough to ignore. Therefore,

$$I_{C(sat)} \cong \frac{V_{CEQ}}{r_E} = \frac{20 \text{ V}}{10 \text{ } \Omega} = 2 \text{ A}$$

189

Transistor Circuit Approximations

The maximum ac output power is

$$P_O = \frac{V_{CEQ}I_{C(sat)}}{2} = \frac{20\text{ V} \times 2\text{ A}}{2} = 20\text{ W}$$

Each transistor dissipates

$$P_D = \frac{P_O}{5} = \frac{20\text{ W}}{5} = 4\text{ W}$$

So, the power rating of each transistor must be at least 4 W at the highest operating temperature.

10-5 A COMPLETE AMPLIFIER

Figure 10-10 is an example of a complete amplifier. It has three stages: a small-signal class A amplifier (Q_1), a large-signal class A amplifier (Q_2), and a class B push-pull emitter follower. The approximate dc voltages of all nodes are listed. The Q_2 stage is called the *driver stage* because it supplies the drive for the output stage. As you see, Q_2 and its emitter resistor have replaced the biasing resistor shown earlier. This direct-coupled design is the most efficient way to connect a driver stage to the output stage.

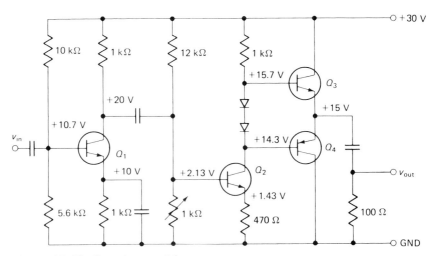

Figure 10-10 Complete amplifier.

EXAMPLE 10-5

Calculate the maximum P_O and P_D for the output stage of Fig. 10-10.

SOLUTION

$$V_{CEQ} = 15 \text{ V}$$

and
$$I_{C(\text{sat})} \cong \frac{V_{CEQ}}{r_E} = \frac{15 \text{ V}}{100 \text{ } \Omega} = 150 \text{ mA}$$

So, the maximum ac output power is

$$P_O \frac{V_{CEQ}I_{C(\text{sat})}}{2} = \frac{15 \text{ V} \times 150 \text{ mA}}{2} = 1.13 \text{ W}$$

The maximum power dissipation of each transistor is

$$P_D = \frac{P_O}{5} = \frac{1.13 \text{ W}}{5} = 226 \text{ mW}$$

EXAMPLE 10-6

What does I_{CQ} equal in the output stage of Fig. 10-10?

SOLUTION

The emitter resistor of the driver stage has 1.43 V across it. Therefore, the emitter current in Q_2 is

$$I_E = \frac{1.43 \text{ V}}{470 \text{ } \Omega} = 3.04 \text{ mA}$$

This 3.04 mA flows through the diodes and the upper 1-kΩ resistor. As a result, the current mirror produces a quiescent collector current through the output transistors of

$$I_{CQ} \cong 3.04 \text{ mA}$$

10-6 THE TUNED CLASS C AMPLIFIER

Figure 10-11a is an example of a class C amplifier. The input coupling capacitor, the base resistor, and the emitter diode function as a negative clamper, so that only the peaks of the ac input signal produce base current.

191

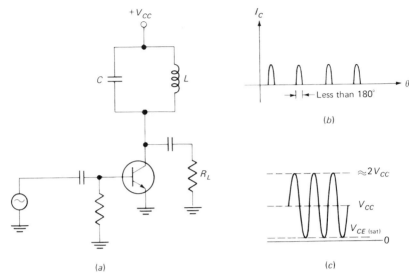

Figure 10-11 Class C amplifier. *(a)* Circuit. *(b)* Current pulses. *(c)* Output voltage.

As a result, the collector current is a train of narrow pulses, similar to Fig. 10-11*b*. This nonsinusoidal signal contains a *fundamental* frequency plus *harmonics.*

When the narrow current pulses drive the resonant collector circuit, all higher harmonics are filtered out, leaving only the fundamental frequency. In other words, the voltage at the collector is sinusoidal (see Fig. 10-11*c*). Since the quiescent collector voltage is V_{CC}, the maximum signal swing is from approximately $V_{CE(sat)}$ to $2V_{CC}$.

The class C amplifier is more efficient than class A or B because it can deliver more ac output power for the same transistor dissipation. But it has to use a resonant load, and for this reason class C amplifiers have limited use.

10-7 SWITCHING OPERATION

A transistor is often used as a switch to control large load currents in industrial applications. Figure 10-12*a* illustrates the idea. The transistor is unbiased, or normally off. This means the quiescent point is at cutoff in Fig. 10-12*b*. In this case, no current exists in load resistor R_L, which may be a lamp, relay, or other heavy load.

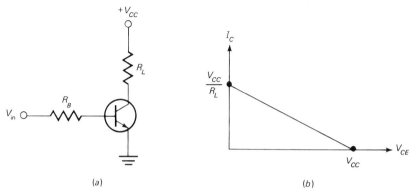

(a)   (b)

Figure 10-12 Switching operation. (a) Circuit. (b) Ends of load line.

When V_{IN} is large enough, the operating point switches from cutoff to saturation and the load current is approximately

$$I_{C(\text{sat})} \cong \frac{V_{CC}}{R_L} \qquad (10\text{-}8)$$

The input voltage that just produces saturation is

$$V_{IN(\text{sat})} = I_{B(\text{sat})}R_B + V_{BE} \qquad (10\text{-}9)$$

where $I_{B(\text{sat})} = I_{C(\text{sat})}/\beta_{\text{dc}}$. As long as the input voltage is greater than the V_{IN} given by Eq. (10-9), the transistor operates as a switch.

Here are a few of the important quantities listed on the data sheet of a power switching transistor:

I_{CEO} = cutoff current with base open
I_{CES} = cutoff current with base shorted to emitter
$V_{CE(\text{sat})}$ = collector-emitter voltage at saturation
$V_{BE(\text{sat})}$ = base-emitter voltage at saturation
t_{on} = time to switch from cutoff to saturation
t_{off} = time to switch from saturation to cutoff

As an example, a 2N5939 has the following values for $I_{C(\text{sat})} = 5$ A and $V_{CE(\text{cutoff})} = 50$ V: $I_{CEO} = 2$ mA, $I_{CES} = 2$ mA, $V_{CE(\text{sat})} = 0.6$ V, $V_{BE(\text{sat})} = 1.2$ V, $t_{\text{on}} = 135$ ns, and $t_{\text{off}} = 800$ ns. Notice that it takes more time to turn a transistor off than it does to turn it on. This is because of charge storage, described earlier in Chap. 3.

Transistor Circuit Approximations

EXAMPLE 10-7

The 2N5939 has a $V_{CE(\text{sat})}$ of 1 V maximum when $I_{C(\text{sat})} = 10$ A. Work out the exact value of load current when the transistor is saturated in Fig. 10-13a. Also, what is the base current that just produces saturation if $h_{FE} = 90$?

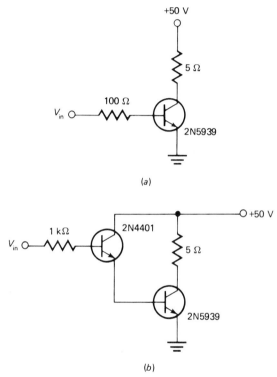

(a)

(b)

Figure 10-13

SOLUTION

The exact saturation current is

$$I_{C(\text{sat})} = \frac{V_{CC} - V_{CE(\text{sat})}}{R_L} = \frac{50\ \text{V} - 1\ \text{V}}{5\ \Omega} = 9.8\ \text{A}$$

The corresponding base current is

$$I_{B(\text{sat})} = \frac{I_{C(\text{sat})}}{\beta\text{dc}} = \frac{9.8\ \text{A}}{90} = 109\ \text{mA}$$

194

EXAMPLE 10-8

In the preceding example, it takes 109 mA of base current to control 9.8 A of collector current. By using an emitter follower as shown in Fig. 10-13b, we can reduce the required input current. The 2N4401 has an h_{FE} of 200. Calculate the values of $I_{B(\text{sat})}$ and $V_{\text{IN(sat)}}$, using 0.7 V and 1.6 V for the V_{BE} drops.

SOLUTION

As calculated in the preceding example, it takes 109 mA of base current to saturate the 2N5939. The $I_{B(\text{sat})}$ of the 2N4401 is

$$I_{B(\text{sat})} = \frac{109 \text{ mA}}{200} = 0.545 \text{ mA}$$

Because of the current gain of the emitter follower, it now takes only 0.545 mA to control 9.8 A of load current. The required input voltage is

$$
\begin{aligned}
V_{\text{IN(sat)}} &= I_{B(\text{sat})}R_B + V_{BE(1)} + V_{BE(2)} \\
&= (0.545 \text{ mA}) (1 \text{ k}\Omega) + 0.7 \text{ V} + 1.6 \text{ V} \\
&= 2.85 \text{ V}
\end{aligned}
$$

GLOSSARY

compensating diode In a class B push-pull amplifier, this is a diode whose curve matches the transconductance curve of the emitter diode.

crossover distortion When unbiased, a class B push-pull amplifier has no output until the input voltage exceeds approximately 0.7 V. This results in clipping between the positive and negative half cycles. To eliminate crossover distortion, a slight forward bias must be applied to the emitter diodes.

current mirror A circuit with a compensating diode in parallel with the emitter diode. When the compensating diode is matched to the emitter diode, the collector current equals the current through the compensating diode.

driver stage The stage supplying the signal drive for a class B push-pull emitter follower.

harmonics Any nonsinusoidal signal is the superposition of sinusoidal signals with frequencies of f, $2f$, $3f$, . . . , nf. Frequency f is called the fundamental or first harmonic, $2f$ is the second harmonic, $3f$ is the third harmonic, and so on. Frequency f equals $1/T$, where T is the period of the nonsinusoidal signal.

switching operation Rather than swing the operating point sinusoidally over the ac load line, switching operation means forcing the operating point to jump suddenly from cutoff to saturation, or vice versa. In this type of operation, the transistor functions as a switch, either floating the load resistor or grounding it.

switching time The jump from cutoff to saturation, or vice versa, is fast but not instantaneous. It takes time to turn on a transistor (t_{on}) and more time to turn it off (t_{off}). Because of charge storage, t_{off} is always greater than t_{on}.

SELF-TESTING REVIEW

Read each of the following and provide the missing words. Answers appear at the beginning of the next question.

1. The transistor of a class A amplifier operates in the active region throughout the ac cycle. This means collector current flows for _____. In a class B circuit, current exists for only _____. In a class C circuit, it flows for less than _____.

2. (*360°, 180°, 180°*) The current mirror is widely used in linear integrated circuits. A _____ diode is in parallel with the emitter diode. The current through this _____ diode produces an equal collector current.

3. (*compensating, compensating*) The transistors in a class B push-pull amplifier are usually _____, meaning similar transconductance curves, maximum ratings, etc.

4. (*complementary*) The main method for biasing a class B push-pull amplifier is _____ bias. The transistors are complementary, and the curves of the compensating diodes _____ the transconductance curves of the transistors. This type of bias is ideal for linear integrated circuits.

5. (*current-mirror, match*) In a class B push-pull amplifier, the quiescent collector current should be slightly greater than zero to avoid _____ distortion. As a guide, an I_{CQ} from 1 to 5 percent of $I_{C(sat)}$ eliminates this type of distortion.

6. (*crossover*) To increase the input impedance, we can use Darlington pairs in a class B push-pull amplifier. Alternatively, we can use complementary pairs. The Darlington pairs require _____ compensating diodes; the complementary pairs need only _____.

196

7. *(four, two)* In a class C amplifier, the collector current is a train of narrow _____. This nonsinusoidal current contains a fundamental frequency plus harmonics. Because of the resonant collector circuit, all higher harmonics are filtered, leaving only the _____ frequency. This means the collector voltage is sinusoidal.

8. *(pulses, fundamental)* A transistor is often used as a switch to control large load currents in industrial applications. When V_{IN} is large enough, the operating point switches from _____ to _____ on the load line.

9. *(cutoff, saturation)* Some of the important quantities listed on the data sheet of a power switching transistor are I_{CEO}, I_{CES}, $V_{CE(sat)}$, $V_{BE(sat)}$, t_{on}, and t_{off}.

PROBLEMS

10-1. What is the class of operation for each of the following collector currents:

a. Exists for $210°$
b. For $180°$
c. For $360°$
d. For $30°$

10-2. What is the collector current for the *npn* mirror of Fig. 10-14a?

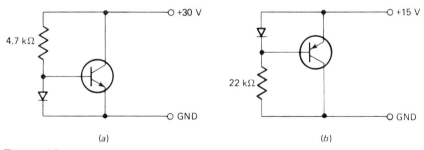

(a)   (b)

Figure 10-14

10-3. In Fig. 10-14b, what is the collector current, assuming perfect mirror action?

10-4. What is the quiescent collector current in Fig. 10-15?

Transistor Circuit Approximations

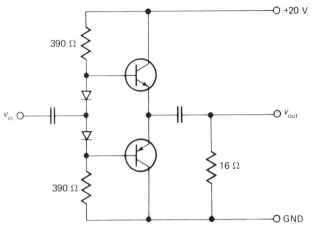

Figure 10-15

10-5. If the supply voltage changes from 20 to 30 V in Fig. 10-15, what does the quiescent collector current equal?

10-6. What does $I_{C(\text{sat})}$ equal in Fig. 10-15?

10-7. Calculate the maximum ac output power in Fig. 10-15. What is the required power rating of each transistor?

10-8. If the supply voltage of Fig. 10-15 changes from 20 to 30 V, what is the maximum ac output power? The required power rating of the transistors?

10-9. In Fig. 10-15, $\beta_{\text{dc}} = 50$ and $R'_e = 0.2\ \Omega$. Ignoring the impedance of the compensating diodes, what does $z_{\text{in(base)}}$ equal? Voltage gain A? Power gain G?

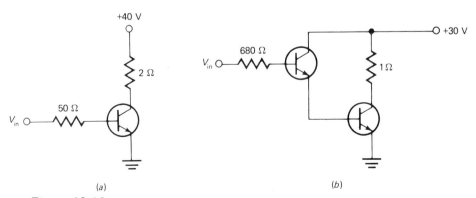

(a)   (b)

Figure 10-16

198

10-10. The transistor of Fig. 10-16a has a V_{BE} drop of 1.1 V at saturation. Ignoring $V_{CE(sat)}$, what is the maximum current through the load resistor? If $\beta_{dc} = 80$, what is the input voltage that just produces saturation?

10-11. If $V_{CE(sat)} = 1.2$ V in Fig. 10-16a, what is the exact value of collector current when the transistor is saturated?

10-12. The input transistor of Fig. 10-16b has an h_{FE} of 250, and the output transistor has an h_{FE} of 75. Ignoring $V_{CE(sat)}$, what is the maximum load current? What is the input base current that just produces saturation? If the input and output transistors have V_{BE} drops of 0.7 V and 1.2 V, what is the input voltage that just produces maximum load current?

Chapter 11

Cascading Stages, Frequency Response, and *h* Parameters

There are three major topics in this chapter. First, we discuss ways to connect transistor stages, including direct coupling. Second, the effect of frequency is covered. Third, the *h*-parameter method of analysis is examined.

11-1 RC COUPLING

Figure 11-1 illustrates *resistance-capacitance* (RC) coupling, the method most widely used in discrete circuits. In this approach, the signal developed across the collector resistor of each stage is coupled into the base of the next stage. In this way, the *cascaded* (one after another) stages amplify the signal, and the overall gain equals the product of the individual gains.

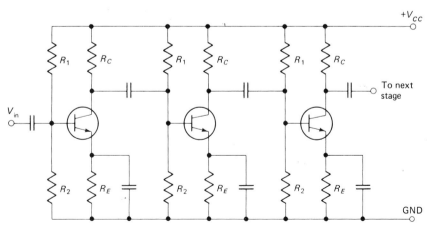

Figure 11-1 RC-coupled stages.

200

The coupling capacitors transmit ac but block dc. Because of this, the stages are isolated as far as dc is concerned. This is necessary to prevent dc interference and shifting of Q points. The drawback of this approach is the drop in voltage gain when the frequency is too low. In other words, as the frequency decreases, the capacitive reactance increases until it begins to block some of the ac signal.

The bypass capacitors are needed because they bypass the emitters to ac ground. Without them, the voltage gain of each stage would be lost. These bypass capacitors also place a lower limit on the frequency of operation. At low enough frequencies, capacitive reactance is high and the ac ground on the emitter is lost.

If you are interested in amplifying ac signals with frequencies greater than 10 Hz, the RC-coupled amplifier is suitable. For discrete circuits it is the most convenient and least expensive way to build a multistage amplifier.

11-2 TWO-STAGE FEEDBACK

Earlier, we discussed single-stage feedback using a swamping resistor. It is also possible to use feedback around two stages as shown in Fig. 11-2. The input signal v_{in} is amplified and inverted in the first stage. The output of this first stage is amplified and inverted again by the second stage. Part of this second-stage output is fed back to the first stage via the voltage divider

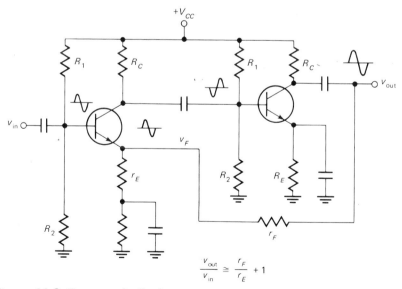

$$\frac{v_{out}}{v_{in}} \cong \frac{r_F}{r_E} + 1$$

Figure 11-2 Two-stage feedback.

formed by r_F and r_E. The feedback signal v_F is applied to the emitter of the first stage. Because this signal returns in phase with the input signal, the base-emitter voltage decreases. This is called *negative feedback*.

Basic idea

As will be proved in a later chapter, negative feedback reduces the overall voltage gain. In exchange for this loss of gain, we get gain stability. Here is the idea. Suppose the r_e' of the second stage becomes smaller. The output signal will increase; however, more signal is fed back to the emitter of the first transistor, reducing the base-emitter voltage. This results in less output signal from the first stage, which almost offsets the original increase in output voltage.

Similarly, if the r_e' of the second stage increases, we get less output signal. This means less feedback signal to the first stage. The base-emitter voltage of the first stage then increases, producing a larger signal out of the first stage. With more input to the second stage, we get more output signal, which almost offsets the original decrease.

Voltage gain

With methods discussed in a later chapter, we can prove that the overall voltage gain is approximately given by

$$A \cong \frac{r_F}{r_E} + 1 \qquad \textbf{(11-1}a\textbf{)}$$

This tells us the overall voltage gain depends on the ratio of two discrete resistors. If precision resistors are used, we get a precise value of voltage gain. Furthermore, it means you can replace transistors without changing the voltage gain. This is ideal for mass production: fixed voltage gain that is independent of transistors and temperature.

Open-loop and closed-loop gain

If we open the feedback resistor r_F in Fig. 11-2, the voltage gain will increase sharply because the negative feedback is lost. The gain with the feedback loop open is called the *open-loop gain;* it's designated A_{OL}. In Fig. 11-2,

$$A_{OL} = A_1 A_2$$

where $A_1 \cong r_{C1}/r_{e1}'$ and $A_2 \cong r_{C2}/r_{e2}'$. If each stage has a voltage gain of 100, the open-loop gain is

$$A_{OL} = 100 \times 100 = 10{,}000$$

The gain with the feedback loop closed is called the *closed-loop gain*, symbolized A_{CL}. In other words, Eq. (11-1*a*) is equivalent to

$$A_{CL} \cong \frac{r_F}{r_E} + 1 \qquad\qquad \textbf{(11-1}b\textbf{)}$$

In a well-designed feedback circuit, the closed-loop gain is much smaller than the open-loop gain. For instance, if r_F is 4.7 kΩ and r_E is 100 Ω, the gain of Fig. 11-2 is

$$A_{CL} \cong \frac{4700}{100} + 1 = 48$$

Chapter 14 will tell you more about open-loop and closed-loop gain.

EXAMPLE 11-1

What is the voltage gain in Fig. 11-2 if $r_F = 9.9$ kΩ and $r_E = 100$ Ω?

SOLUTION

$$A \cong \frac{r_F}{r_E} + 1 = \frac{9.9 \text{ k}\Omega}{100 \text{ }\Omega} + 1 = 100$$

11-3 INDUCTIVE COUPLING

Occasionally, transistor stages will use *RF chokes* (inductors at radio frequencies) instead of collector resistors, as shown in Fig. 11-3. The idea is to

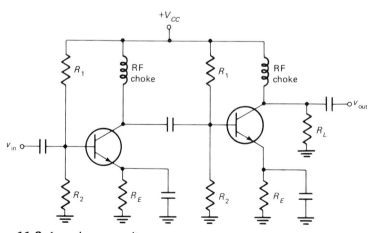

Figure 11-3 Impedance coupling.

prevent the loss of signal power that takes place in resistors. Such amplifiers are inherently *high-pass* amplifiers; this means they are intended to amplify frequencies that are high enough for the RF chokes to appear as ac open circuits. In this case, the ac load resistance seen by the first stage is the input impedance of the second stage. The ac load resistance seen by the second stage is the final load resistance R_L.

11-4 TRANSFORMER COUPLING

Figure 11-4 shows *transformer coupling.* Small ac load resistances can be stepped up to higher impedance levels. This improves voltage gain. Furthermore, since no signal power is wasted in a collector resistor, all the ac power is delivered to the final load.

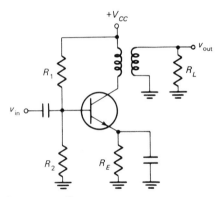

Figure 11-4 Transformer coupling.

Transformer coupling was once popular at *audio frequencies* (20 Hz to 20 kHz). But the cost and bulkiness of transformers that operate at audio frequencies is a major disadvantage. With the advent of complementary transistors, the class B push-pull emitter follower soon replaced transformer coupling in most audio amplifiers.

The one area where transformer coupling has survived is *radio frequency* (RF) amplifiers. Radio frequency means all frequencies greater than 20 kHz. In AM radio receivers, RF ranges from 550 to 1600 kHz. In TV receivers, the RF signals have frequencies from 54 to 216 MHz (channels 2 through 13). Transformer coupling is still used in RF amplifiers because the transformers are much smaller and less expensive than audio transformers.

204

11-5 TUNED TRANSFORMER COUPLING

Often, the stages of an RF amplifier are designed to amplify a narrow band of frequencies. This is how a radio or TV receiver separates one station from another. Figure 11-5 is an example of tuned transformer coupling.

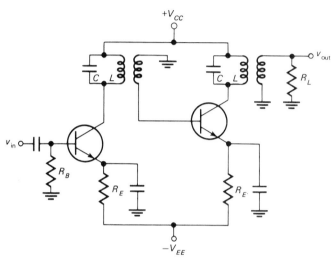

Figure 11-5 Tuned-transformer coupling.

At resonance, the impedance of each LC tank is high. Above and below the resonant frequency the impedance decreases. Therefore, the voltage gain is maximum at resonance.

The formula for the resonant frequency is

$$f_r \cong \frac{1}{2\pi\sqrt{LC}} \qquad\qquad (11\text{-}2)$$

The bandwidth is

$$B = \frac{f_r}{Q} \qquad\qquad (11\text{-}3)$$

where Q is the ratio of the ac collector resistance to the inductive reactance of the primary winding:

$$Q = \frac{r_C}{X_L} \qquad \textbf{(11-4)}$$

Each collector works into a parallel LC tank circuit. At resonance, the inductive reactance cancels the capacitive reactance, leaving a purely resistive load on each collector. Above resonance, X_C becomes smaller than X_L, causing the voltage gain to decrease. Below resonance, X_L is smaller than X_C, and the voltage gain again drops off.

A tuned transformer coupling like Fig. 11-5 produces a *bandpass amplifier*, one with maximum gain at the resonant frequency. As we move away from the resonant frequency, the voltage gain drops off rapidly. As mentioned earlier, this *selectivity* is what allows a receiver to separate one station from another.

11-6 DIRECT COUPLING

All amplifiers discussed so far have been capacitor-coupled or transformer-coupled between stages. This limits the low-frequency response. In other words, the gain of the amplifiers drops off at lower frequencies. One way to avoid this is with *direct coupling;* this means providing a dc path between stages.

One-supply circuit

Figure 11-6 is a two-stage direct-coupled amplifier; no coupling or bypass capacitors are used. Because of this, dc is amplified as well as ac. With a

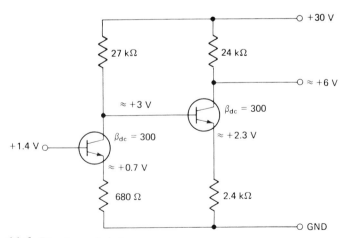

Figure 11-6 Direct-coupled stages.

206

quiescent input voltage of +1.4 V, about 0.7 V is dropped across the first emitter diode, leaving +0.7 V across the 680-Ω resistor. This sets up approximately 1 mA of collector current. This 1 mA then produces a 27-V drop across the collector resistor. Therefore, the first collector runs at about +3 V with respect to ground.

Allowing 0.7 V for the second emitter diode, we get 2.3 V across the 2.4-kΩ emitter resistor. This results in approximately 1 mA of collector current, about a 24-V drop across the collector resistor, and around +6 V from the second collector to ground. Therefore, a quiescent input voltage of +1.4 V gives a quiescent output voltage of +6 V.

Because of the high β_{dc}, we can ignore the loading effect of the second base upon the first collector. Ignoring r'_e, the first stage has a voltage gain of

$$A_1 = \frac{27{,}000}{680} \cong 40$$

The second stage has a voltage gain of

$$A_2 = \frac{24{,}000}{2400} = 10$$

And the overall gain is

$$A = A_1 A_2 = 40 \times 10 = 400$$

The two-stage circuit will amplify any change in the input voltage by a factor of 400. For instance, if the input voltage changes by +5 mV, the final output voltage changes by

$$\Delta v_{out} = 400 \times 5 \text{ mV} = 2 \text{ V}$$

Because two inverting stages are involved, the final output goes from +6 to +8 V.

Here is the main disadvantage of direct coupling. Transistor characteristics like V_{BE} change with temperature. This causes quiescent collector currents and voltages to change. Because of the direct coupling, the voltage changes are coupled from one stage to the next, appearing at the final output as an amplified voltage change. This unwanted change is called *drift*. The trouble with drift is you can't distinguish it from a genuine change produced by the input signal.

207

Ground-referenced input

For the two-stage amplifier of Fig. 11-6 to work properly, we need a quiescent input voltage of +1.4 V. In typical applications, it's necessary to have a *ground-referenced* input, one where the quiescent input voltage is 0 V.

Figure 11-7 shows a ground-referenced input stage. This stage is a *pnp* Darlington with the input base returned to ground through the signal source. Because of this, the first emitter is approximately +0.7 V above ground, and the second emitter is about +1.4 V above ground. The +1.4 V biases the second stage, which operates as previously described.

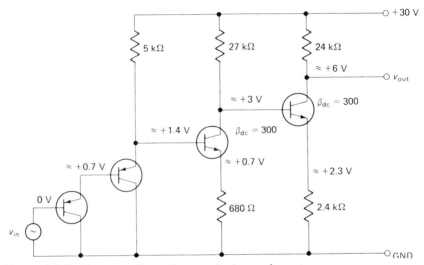

Figure 11-7 Direct coupling with ground-referenced input.

The quiescent V_{CE} of the first transistor is only 0.7 V, and the quiescent V_{CE} of the second transistor is only 1.4 V. Nevertheless, both transistors are operating in the active region because the $V_{CE(sat)}$ of low-power transistors is only about 0.1 V. Since the input signal is typically in millivolts, the input transistors continue to operate in the active region.

Remember the *pnp* ground-referenced input. It is used a lot in audio integrated circuits.

Two-supply circuit

When a *split supply* (both positive and negative voltages) is available, we can reference both the input and the output to ground. Figure 11-8 is an example. The first stage is emitter-biased with an I_E around 1 mA. This

208

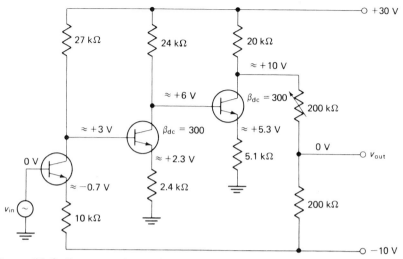

Figure 11-8 Direct coupling with ground-referenced input and output.

produces about +3 V at the first collector. Subtracting the V_{BE} drop of the second emitter diode leaves +2.3 V at the second emitter.

The emitter current in the second stage is around 1 mA. This flows through the collector resistor, producing about +6 V from the collector to ground. The final stage has +5.3 V across the emitter resistor, which gives about 1 mA of current. Therefore, the last collector has approximately +10 V to ground.

The output voltage divider references the output to ground. When the upper resistor is adjusted to 200 kΩ, the final output voltage is approximately 0 V. The adjustment allows us to eliminate errors caused by resistor tolerances, V_{BE} differences from one transistor to the next, etc.

What is the overall voltage gain? The first stage has a gain around 2.7, the second stage around 10, the third stage around 4, and the voltage divider about 0.5. Therefore,

$$A = 2.7 \times 10 \times 4 \times 0.5 = 54$$

Review of direct coupling

This gives you the idea behind direct coupling. Coupling and bypass capacitors are not used, allowing dc as well as ac to reach the next stage. As a result, the amplifier has no lower frequency limit; it amplifies all frequencies down to zero. Herein lie the strength and the weakness of direct coupling: it's good to be able to amplify very low frequencies, including dc; it's bad,

209

however, to amplify very slow changes in supply voltage, transistor variations, etc.

There is a way to reduce the drift with a special two-transistor stage called a *differential* amplifier. The next section discusses this widely used amplifier.

11-7 THE DIFFERENTIAL AMPLIFIER

The *differential amplifier* (diff amp) is extensively used in linear ICs. A diff amp has no coupling or bypass capacitors; all it requires are resistors and transistors, both easily fabricated on a chip.

Tail current

Figure 11-9 shows a diff amp. There are two input signals and one output signal. The output is the voltage between the collectors. The circuit is symmetrical; ideally, each half is identical to the other half. Integrated circuits can approach this symmetry because the components are on the same chip and have almost identical characteristics.

A diff amp is sometimes called a *long-tail pair* because it consists of a pair of identical transistors connected to a common emitter resistor (the tail). The current through this common resistor is called the *tail current, I_T.*

The diff amp of Fig. 11-9 uses emitter bias, discussed earlier. As you recall, the key to analyzing emitter-biased circuits is to remember that the

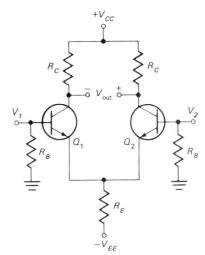

Figure 11-9 Differential amplifier.

210

top of the emitter resistor is an approximate ground point. Because of this, almost all the V_{EE} supply voltage appears across R_E, and

$$I_T \cong \frac{V_{EE}}{R_E} \tag{11-5}$$

The tail current divides equally between the two transistors, so that each has an emitter current of

$$I_E = \frac{I_T}{2} \tag{11-6}$$

AC operation

A diff amp acts like a bridge circuit. Voltages v_1 and v_2 control the currents in each transistor. If v_1 and v_2 are equal, the bridge is balanced and the output is zero. On the other hand, when v_1 and v_2 are unequal, the bridge is unbalanced and we get an output.

From the v_1 input side, the diff amp appears to be an emitter follower driving a common-base amplifier. In other words, v_1 drives Q_1, which acts like an emitter follower that couples the signal to the base of Q_2. Then Q_2 acts like a CB amplifier. By analyzing the ac equivalent circuit, we can come up with the following formula:

$$v_{\text{out}} = A(v_1 - v_2) \tag{11-7}$$

where $A \cong R_C/r_e'$. This means the circuit amplifies the difference of the two inputs. For example, if $R_C = 10 \text{ k}\Omega$ and $r_e' = 50 \text{ }\Omega$, $A = 200$. The output voltage will equal 200 times the difference of the input voltages.

Since Q_2 acts like a common-base amplifier, it has an input impedance of r_e'. This r_e' is in series with the r_e' of Q_1; therefore, the input impedance looking into the base of Q_1 is

$$z_{\text{in(base)}} \cong \beta(r_E + r_e') = \beta(r_e' + r_e')$$

or
$$z_{\text{in(base)}} \cong 2\beta r_e' \tag{11-8}$$

From the v_2 input side, the action is complementary. Q_2 acts like an emitter follower driving Q_1, which acts like a common-base amplifier. Therefore, the input impedance looking into the base of Q_2 is given by Eq. (11-8).

211

Transistor Circuit Approximations

Single-ended input

There is no need to use both inputs. One of the inputs can be zero, as shown in Fig. 11-10. Since Q_1 acts like an emitter follower, the signal driving the Q_2 emitter is in phase with the v_1 signal. Notice the phases of the collector signals. The Q_1 collector signal is out of phase with the v_1 signal, and the Q_2 collector signal is in phase. Voltage v_{out} is the algebraic difference of these two collector signals.

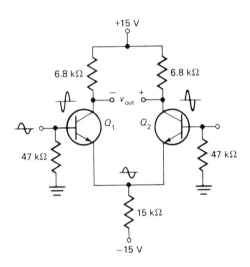

Figure 11-10

EXAMPLE 11-2

Calculate the voltage gain and input impedance in Fig. 11-10 for a β of 150.

SOLUTION

First, get the tail current:

$$I_T \cong \frac{15 \text{ V}}{15 \text{ k}\Omega} = 1 \text{ mA}$$

which means each emitter gets

$$I_E = \frac{I_T}{2} = \frac{1 \text{ mA}}{2} = 0.5 \text{ mA}$$

212

So, the r'_e of each transistor is ideally 50 Ω. The voltage gain is

$$A \cong \frac{6.8 \text{ k}\Omega}{50 \text{ }\Omega} = 136$$

and the input impedance of the base is

$$z_{\text{in(base)}} \cong 2\beta r'_e = 2 \times 150 \times 50 \text{ }\Omega = 15 \text{ k}\Omega$$

11-8 RESPONSE OF AN RC-COUPLED AMPLIFIER

The RC-coupled amplifier is the most common type of discrete amplifier. It uses resistors to develop the ac signal and capacitors to couple and bypass it. A typical response for a fixed input voltage is shown in Fig. 11-11. For very low or very high frequencies, the output voltage drops off. Toward the middle of the frequency range, however, the output voltage is fixed and equals a constant value of K. It is in this middle range of frequencies that RC-coupled amplifiers normally operate.

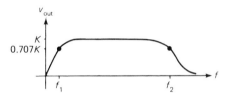

Figure 11-11 Amplifier passband.

Cutoff frequencies

In Fig. 11-11, the output voltage equals K in the mid-frequency range. If we increase or decrease the frequency, we reach a point where the output voltage drops to $0.707K$. The frequencies for these $0.707K$ points are known as the *cutoff frequencies* (sometimes called the *break frequencies*).

Figure 11-11 shows two cutoff frequencies, f_1 and f_2. Coupling and bypass capacitors are responsible for the lower cutoff frequency f_1. Stray and internal transistor capacitances are the cause of the upper cutoff frequency f_2. All frequencies between f_1 and f_2 are the *passband* of the amplifier. Outside the passband, the gain drops off rapidly.

Bandwidth

The *bandwidth, B,* is important when discussing amplifiers. It's defined as the width of the passband:

213

Transistor Circuit Approximations

$$B = f_2 - f_1$$

For instance, if $f_1 = 10$ Hz and $f_2 = 50$ kHz, then

$$B = 50,000 \text{ Hz} - 10 \text{ Hz} \cong 50 \text{ kHz}$$

In a direct-coupled amplifier, f_1 is zero and B equals f_2. This applies to *op amps,* direct-coupled amplifiers discussed in Chap. 14.

11-9 LOWER CE CUTOFF FREQUENCIES

In a CE amplifier like Fig. 11-12, the coupling and bypass capacitors produce lower cutoff frequencies. What happens is this: As the frequency decreases, the capacitive reactance increases. Eventually, the coupling capacitors drop a significant amount of the signal, and the bypass capacitor no longer ac-grounds the emitter.

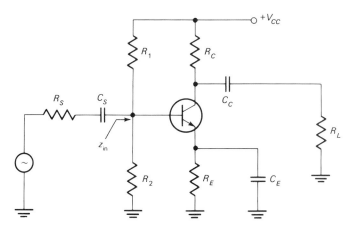

Figure 11-12 CE amplifier.

Input coupling capacitor

In Fig. 11-12, the input impedance of the stage is

$$z_{\text{in}} = R_1 \| R_2 \| z_{\text{in(base)}}$$

The input coupling capacitor is in series with R_S and z_{in}. As discussed in basic circuit theory, the ac current is down to the 0.707 point when the capacitive reactance equals the total series resistance:

214

$$X_C = R$$

or
$$\frac{1}{2\pi f C_S} = R_S + z_{in}$$

Solving for frequency, we get

$$f = \frac{1}{2\pi(R_S + z_{in})C_S} \qquad (11\text{-}9)$$

This is the cutoff frequency produced by the input coupling capacitor.

Output coupling capacitor

On the output side, coupling capacitor C_C is in series with R_C and R_L. The output ac current is down to the 0.707 point when

$$X_C = R$$

or
$$\frac{1}{2\pi f C_C} = R_C + R_L$$

Solving for frequency gives

$$f = \frac{1}{2\pi(R_C + R_L)C_C} \qquad (11\text{-}10)$$

This is the cutoff frequency produced by the output coupling capacitor.

Emitter bypass capacitor

In Fig. 11-12, the capacitor sees a resistance of R_E in parallel with the output impedance of the emitter. This output impedance equals the ac emitter resistance and the base resistance divided by the current gain. The formula is

$$z_{out} = r'_e + \frac{R_S \parallel R_1 \parallel R_2}{\beta}$$

This is the impedance looking back into the emitter. The equivalent resistance in parallel with the bypass capacitor is $R_E \parallel z_{out}$.

215

Transistor Circuit Approximations

When the capacitive reactance equals the total parallel resistance, the voltage gain drops to the 0.707 point. In other words, the cutoff frequency occurs when

$$X_C = R$$

or
$$\frac{1}{2\pi f C_E} = R_E \parallel z_{\text{out}}$$

Solving for frequency, we get

$$f = \frac{1}{2\pi (R_E \parallel z_{\text{out}}) C_E} \tag{11-11}$$

This is the cutoff frequency produced by the emitter bypass capacitor.

Which one to use

The three capacitors produce three different cutoff frequencies. The highest one is the critical one because that's where the voltage gain first starts to drop off. For instance, if the three cutoff frequencies are 10 Hz, 50 Hz, and 200 Hz, then the 200 Hz is the most important. Usually, the cutoff frequencies are different; only by a coincidence would they all be the same. Therefore, when analyzing a CE amplifier, the largest of the lower cutoff frequencies is the limiting frequency.

11-10 TRANSISTOR CUTOFF FREQUENCIES

As the frequency increases, certain things happen inside a transistor that reduce the current and voltage gain. Bear in mind that a transistor has internal capacitances, charge storage, and other effects that can alter the high-frequency response.

Alpha cutoff frequency

The ac alpha of a transistor is the ratio of ac collector current to ac emitter current:

$$\alpha = \frac{i_c}{i_e}$$

216

At low frequencies, α approaches unity. But as the frequency increases, we eventually reach the point where charge storage has an effect. And the effect is a drop-off in the value of α.

The *alpha cutoff frequency* f_α is the frequency where α has dropped to 0.707 of its low-frequency value. For instance, if a transistor has an α of 0.98 at low frequencies and an f_α of 300 MHz, then at a frequency of 300 MHz,

$$\alpha = 0.707 \times 0.98 = 0.693$$

This means that at 300 MHz, the ac collector current is only 0.693 times the ac emitter current.

The f_α represents one of the limitations on the upper-frequency response of a CB amplifier. When possible, we select a transistor whose f_α is much higher than the highest operating frequency of the CB amplifier.

Beta cutoff frequency

The *beta cutoff frequency* f_β is another important transistor characteristic. It is the frequency where the β of the transistor has decreased to 0.707 of its low-frequency value. For instance, if a transistor has a low-frequency β of 250 and an f_β of 2 MHz, then at 2 MHz the β of the transistor is

$$\beta = 0.707 \times 250 = 177$$

Current-gain bandwidth product

The *current-gain bandwidth product* f_T is another important high-frequency characteristic of a transistor. It is the frequency where the β equals unity. In other words, when the frequency increases, β keeps decreasing until eventually it drops all the way down to unity. The frequency where this happens is designated f_T, the current-gain bandwidth product.

Relationships

The f_T of a transistor is much higher than the f_β. The relation between these two frequencies is

$$f_\beta = \frac{f_T}{\beta} \qquad \textbf{(11-12)}$$

217

Transistor Circuit Approximations

where β is the low-frequency value of β. As an example, suppose a data sheet lists an f_T of 100 MHz and a low-frequency β of 50. Then,

$$f_\beta = \frac{100 \text{ MHz}}{50} = 2 \text{ MHz}$$

The f_α and f_T are also related. As a rough approximation, $f_\alpha \cong f_T$. In actuality, f_T is less than f_α. For simple junction transistors, a better approximation is

$$f_T = \frac{f_\alpha}{1.2} \tag{11-13}$$

The various cutoff frequencies are important in high-frequency analysis of amplifiers. The f_α is one of the limitations on a CB amplifier. The f_β and f_T are limitations on CE amplifiers. The analysis of transistor amplifiers operating at high frequencies is very complicated and is discussed in advanced books. (For a thorough analysis of cutoff frequencies, see Albert P. Malvino, *Electronic Principles,* Second Edition, McGraw-Hill Book Company, New York, 1979, pages 437–442.) As a rough guide, the f_β of a transistor in a CE amplifier should be higher than the highest frequency you're trying to amplify. Likewise, the f_α in a CB amplifier should be much higher than the highest frequency being amplified.

11-11 SECOND-APPROXIMATION FORMULAS

The ideal transistor is all right for preliminary analysis and design. If you need better accuracy, you need to use the second approximation shown in Fig. 11-13. Resistance r_b' is called the *base-spreading resistance.* Its value depends on the doping level of the base and other factors. In rare cases, r_b' may be as high as 1 kΩ. Typically, it is in the range of 50 to 150 Ω. This section includes the effect of r_b' on voltage gain and impedances. The discussion in the remainder of this chapter is taken from Albert P. Malvino,

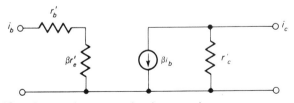

Figure 11-13 AC equivalent circuit for the second approximation.

218

Cascading Stages, Frequency Response, and *h* Parameters

Electronic Principles, Second Edition, McGraw-Hill Book Company, 1979, Chap. 15, pp. 398 ff.

The reverse-biased collector diode has an ac resistance r_c'. This resistance is normally greater than a megohm, and we can expect it to have a small effect on voltage gain and input impedance.

By reanalyzing the CE, CC, and CB amplifiers, we can include the effects of r_b' and r_c' on voltage gain, current gain, input impedance, and output impedance. Table 11-1 summarizes the results.

TABLE 11-1   SECOND-APPROXIMATION FORMULAS

	Common-emitter	Common-collector	Common-base
A_v	$\dfrac{\alpha r_C}{r_e' + r_b'(1-\alpha)}$	$\dfrac{r_E}{r_E + r_e' + r_b'(1-\alpha)}$	$\dfrac{\alpha r_C}{r_e' + r_b'(1-\alpha)}$
A_i	β	$\dfrac{1}{1-\alpha}$	α
z_{in}	$r_b' + \beta r_e'$	$r_b' + \beta(r_E + r_e')$	$r_e' + \dfrac{r_b'}{\beta}$
z_{out}	$\dfrac{r_c'}{\beta} + \dfrac{r_e' r_c'}{R_S + r_b' + r_e'}$	$r_e' + \dfrac{(R_S + r_b')}{\beta}$	$r_c'\left(\dfrac{R_S + r_e'}{R_S + r_b' + r_e'}\right)$

EXAMPLE 11-3

A CE amplifier has the following: $\alpha = 0.99$, $r_C = 2$ kΩ, $r_e' = 25$ Ω, $r_b' = 100$ Ω, $\beta = 100$, $r_c' = 5$ MΩ, and $R_S = 1$ kΩ. Calculate A, A_i, z_{in}, and z_{out}.

SOLUTION

The voltage gain equals

$$A = \frac{\alpha r_C}{r_e' + r_b'(1-\alpha)} = \frac{0.99(2000)}{25 + 100(1 - 0.99)} = 76.2$$

The current gain is

$$A_i = \beta = 100$$

The input impedance equals

$$z_{\text{in}} = r_b' + \beta r_e' = 100 + 100(25) = 2.6 \text{ k}\Omega$$

219

Transistor Circuit Approximations

The output impedance is

$$z_{\text{out}} = \frac{r_c'}{\beta} + \frac{r_e' \, r_c'}{R_S + r_b' + r_e'}$$

$$= \frac{5(10^6)}{100} + \frac{(25)5(10^6)}{1000 + 100 + 25} = 161 \text{ k}\Omega$$

11-12 HYBRID PARAMETERS

Hybrid (h) parameters are easy to measure; this is the reason some data sheets specify low-frequency transistor characteristics in terms of four *h* parameters. This section tells you what *h* parameters are, and how they are related to the *r′* parameters we have been using.

What they are

The four *h* parameters of the CE connection are

h_{ie} = input impedance $(r_C = 0)$
h_{fe} = current gain $(r_C = 0)$
h_{re} = reverse voltage gain $(R_S = \infty)$
h_{oe} = output admittance $(R_S = \infty)$

The first two parameters are specified for an ac load resistance of zero, equivalent to a shorted output. The next two parameters are specified for a source resistance of infinity, equivalent to an open input.

Figure 11-14a shows how to measure h_{ie} and h_{fe}. To begin with, we use an ac short across the output. This prevents loading effects, ensuring

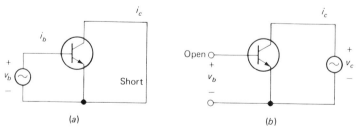

(a)   (b)

Figure 11-14 AC equivalent circuits for measuring *h* parameters. *(a)* Getting h_{ie} and h_{fe}. *(b)* Getting h_{re} and h_{oe}.

unambiguous values for h_{ie} and h_{fe}. In other words, input impedance and current gain will vary if the ac load resistance is too large; to avoid this, we short the output terminals. In Fig. 11-14a, the ratio of input voltage to input current equals the input impedance. In symbols,

$$h_{ie} = \frac{v_b}{i_b} \qquad (r_C = 0)$$

The ratio of output current to input current is the current gain:

$$h_{fe} = \frac{i_c}{i_b} \qquad (r_C = 0)$$

To get the other two parameters, we use the ac equivalent of Fig. 11-14b. Here we drive the collector with a voltage source v_c; this forces a current i_c to flow into the collector. The base is left open; that is, $R_S = \infty$. The ratio of ac base voltage to ac collector voltage is the reverse voltage gain:

$$h_{re} = \frac{v_b}{v_c} \qquad (R_S = \infty)$$

And finally, if we take the ratio of ac collector current to ac collector voltage, we get the output admittance:

$$h_{oe} = \frac{i_c}{v_c} \qquad (R_S = \infty)$$

Relationship to r' parameters

The r' parameters (r'_e, r'_b, r'_c, β, and α) are the easiest to work with; the h parameters are the easiest to measure. Therefore, we need to know how to convert from h parameters to r' parameters. This way, when a data sheet specifies h parameters, we can get r' parameters and analyze transistor amplifiers with the methods of earlier chapters.

Table 11-2 shows the approximate relationships between r' parameters and h parameters. As indicated, $\beta = h_{fe}$, $\alpha = h_{fe}/(h_{fe} + 1)$, and so on. Many data sheets do list the common-emitter h parameters, so with the formulas of Table 11-2 you can convert to r' parameters.

Transistor Circuit Approximations

TABLE 11-2
APPROXIMATE
RELATIONSHIPS

r' Parameter	h Parameter
β	h_{fe}
α	$h_{fe}/(h_{fe}+1)$
r_e'	h_{ie}/h_{fe}
r_c'	h_{fe}/h_{oe}
r_b'	h_{rb}/h_{ob}

The only unusual entry in Table 11-2 is

$$r_b' = \frac{h_{rb}}{h_{ob}}$$

This is the ratio of the reverse voltage gain and output admittance of a common-base circuit. In other words, the easiest and most reliable way to measure r_b' is with a common-base circuit. Reverse voltage gain h_{rb} divided by output admittance h_{ob} gives the value of r_b'.

EXAMPLE 11-4

The data sheet of a 2N3904 shows the following typical values at $I_C = 1$ mA:

$$h_{ie} = 3.5 \text{ k}\Omega$$
$$h_{fe} = 120$$
$$h_{re} = 1.3(10^{-4})$$
$$h_{oe} = 8.5 \text{ }\mu\text{S}$$

Work out the values of β, α, r_e', and r_c'.

SOLUTION

$$\beta = h_{fe} = 120$$

$$\alpha = \frac{h_{fe}}{h_{fe}+1} = \frac{120}{120+1} = 0.992$$

$$r_e' = \frac{h_{ie}}{h_{fe}} = \frac{3500}{120} = 29 \text{ }\Omega$$

$$r_c' = \frac{h_{fe}}{h_{oe}} = \frac{120}{8.5(10^{-6})} = 14.1 \text{ M}\Omega$$

EXAMPLE 11-5

The 2N1975 data sheet specifies $h_{rb} = 1.75(10^{-4})$ and $h_{ob} = 1\ \mu S$ at $I_C = 1$ mA. Calculate r'_b. (Note: S is the symbol for the unit *siemens*, formerly referred to as "mho.")

SOLUTION

$$r'_b = \frac{h_{rb}}{h_{ob}} = \frac{1.75(10^{-4})}{10^{-6}} = 175\ \Omega$$

11-13 APPROXIMATE HYBRID FORMULAS

The *h* parameters are measured with shorted output and open input to avoid the ambiguous values that would result from different source and load resistances. In a practical amplifier, the source resistance is not infinite, and the load resistance is not zero. Nevertheless, *h* parameters can be used to analyze practical amplifiers.

Table 11-3 lists the approximate formulas for the three transistor connec-

TABLE 11-3 APPROXIMATE HYBRID FORMULAS

	Common-Emitter	Common-Collector	Common-Base
A_v	$\dfrac{h_{fe}r_C}{h_{ie}}$	1	$\dfrac{h_{fe}r_C}{h_{ie}}$
A_i	h_{fe}	h_{fe}	1
z_{in}	h_{ie}	$h_{fe}r_E$	$\dfrac{h_{ie}}{h_{fe}}$
z_{out}	$\dfrac{1}{h_{oe}}$	$\dfrac{R_S + h_{ie}}{h_{fe}}$	$\dfrac{h_{fe}}{h_{oe}}$

tions. If you prefer, use these formulas directly instead of converting to r' parameters. These approximate formulas are reasonably accurate for most amplifiers. If you require exact answers, you will need to use the complicated formulas given in the next section.

EXAMPLE 11-6

Using the *h* parameters of Example 11-4, calculate A, A_i, z_{in}, and z_{out} for a CE amplifier with $r_C = 2$ kΩ.

223

Transistor Circuit Approximations

SOLUTION

$$A = \frac{h_{fe} r_C}{h_{ie}} = \frac{120(2000)}{3500} = 68.6$$

$$A_i = h_{fe} = 120$$

$$z_{\text{in}} = h_{ie} = 3.5 \text{ k}\Omega$$

$$z_{\text{out}} = \frac{1}{h_{oe}} = \frac{1}{8.5(10^{-6})} = 118 \text{ k}\Omega$$

EXAMPLE 11-7

The 2N3904 is used as an emitter follower with $r_E = 100 \ \Omega$ and $R_S = 1.5$ kΩ. Work out the values of z_{in} and z_{out}.

SOLUTION

With Table 11-3 and the h parameters of Example 11-4,

$$z_{\text{in}} = h_{fe} r_E = 120(100) = 12 \text{ k}\Omega$$

and $\qquad z_{\text{out}} = \dfrac{R_S + h_{ie}}{h_{fe}} = \dfrac{1500 + 3500}{120} = 41.7 \ \Omega$

EXAMPLE 11-8

The 2N3904 is used in a common-base amplifier. With the h parameters of Example 11-4, calculate z_{in} and z_{out}.

SOLUTION

$$z_{\text{in}} = \frac{h_{ie}}{h_{fe}} = \frac{3500}{120} = 29 \ \Omega$$

and $\qquad z_{\text{out}} = \dfrac{h_{fe}}{h_{oe}} = \dfrac{120}{8.5(10^{-6})} = 14.1 \text{ M}\Omega$

11-14 EXACT HYBRID FORMULAS

The ideal-transistor approximation is adequate for preliminary analysis and design. When more accurate answers are needed, you can use the improved approximations of Tables 11-1 and 11-3. If you need the utmost accuracy, be prepared to use complicated formulas that take everything into account.

224

Numerical parameters

In deriving exact formulas, it's helpful to use numerical parameters h_{11}, h_{12}, h_{21}, and h_{22}. These parameters have the following meaning for any transistor connection:

h_{11} = input impedance   (shorted output)
h_{12} = reverse voltage gain   (open input)
h_{21} = current gain   (shorted output)
h_{22} = output admittance   (open input)

Table 11-4 shows how these numerical parameters are related to the *h* parameters of each transistor connection. For instance, for a CE amplifier

$$h_{11} = h_{ie}$$
$$h_{12} = h_{re}$$
$$h_{21} = h_{fe}$$
$$h_{22} = h_{oe}$$

TABLE 11-4   HYBRID RELATIONSHIPS

Numerical	Common-Emitter	Common-Collector	Common-Base
h_{11}	h_{ie}	h_{ic}	h_{ib}
h_{12}	h_{re}	h_{rc}	h_{rb}
h_{21}	h_{fe}	h_{fc}	h_{fb}
h_{22}	h_{oe}	h_{oc}	h_{ob}

Formulas

Table 11-5 lists the exact formulas for CE, CC, and CB amplifiers. All you have to do is substitute for the numerical parameters. AC load resistance

TABLE 11-5   HYBRID FORMULAS

	Exact	Approximate
A_v	$\dfrac{h_{21}r_L}{h_{11}(1 + h_{22}r_L) - h_{12}h_{21}r_L}$	$\dfrac{h_{21}r_L}{h_{11}}$
A_i	$\dfrac{h_{21}}{1 + h_{22}r_L}$	h_{21}
z_{in}	$h_{11} - \dfrac{h_{12}h_{21}}{h_{22} + 1/r_L}$	h_{11}
z_{out}	$\dfrac{R_S + h_{11}}{(R_S + h_{11})h_{22} - h_{12}h_{21}}$	$\dfrac{1}{h_{22}}$

225

Transistor Circuit Approximations

r_L depends on the transistor connection. With a CE or CB amplifier, $r_L = r_C$. With a CC connection, $r_L = r_E$.

As an example, suppose we want the voltage gain of a common-emitter amplifier. With Tables 11-4 and 11-5,

$$A = \frac{h_{fe}r_C}{h_{ie}(1 + h_{oe}r_C) - h_{re}h_{fe}r_C}$$

Practical comments

The exact formulas of Table 11-5 give exact answers, provided you have the exact h parameters for the transistor being used. This is where a practical problem comes in. The spread in minimum and maximum values is huge. For instance, the data sheet of a 2N3904 gives the following ranges in h parameters for $I_C = 1$ mA:

$$h_{ie} = 1 \text{ to } 10 \text{ k}\Omega$$
$$h_{re} = 0.5(10^{-4}) \text{ to } 8(10^{-4})$$
$$h_{fe} = 100 \text{ to } 400$$
$$h_{oe} = 1 \text{ to } 40 \text{ } \mu S$$

With spreads like these, exact formulas lose their appeal. When working with thousands of 2N3904s, all the formulas give us is an estimate of the voltage gain, input impedance, etc.

Here's the point. All transistors have large tolerances in their specified h parameters. For this reason, we wind up with approximate answers, so we may as well use ideal formulas because they introduce much less error than the huge tolerance of the h parameters. In other words, the ideal formulas give us typical answers, located somewhere between the minimum and maximum answers predicted by exact analysis.

EXAMPLE 11-9

A 2N3904 is used in a CE amplifier with $r_C = 2$ kΩ. With the h parameters of Example 11-4, calculate the exact voltage gain.

SOLUTION

$$A = \frac{h_{fe}r_C}{h_{ie}(1 + h_{oe}r_C) - h_{re}h_{fe}r_C}$$

$$= \frac{120(2000)}{3500 \, [1 + 8.5(10^{-6})2000] - 1.3(10^{-4})120(2000)}$$

$$= 68$$

226

GLOSSARY

audio frequency Any frequency between 20 Hz and 20 kHz.

bandwidth In a tuned amplifier or resonant circuit, this is the difference between the two half-power frequencies.

base-spreading resistance This is the ac resistance of the base region. Designated r_b', its value is typically between 50 and 150 Ω, although it occasionally may be as high as 1 kΩ.

cascade A connection of stages where the output of one stage is the input to the next stage.

cutoff frequency For an amplifier, this is the frequency where the gain is down to 0.707 times its maximum value.

differential amplifier A two-transistor direct-coupled stage that amplifies the difference of two input signals.

direct coupling A circuit where the dc voltage of one stage is coupled to the next stage.

drift In a direct-coupled amplifier this is the change in output voltage caused by temperature, power-supply variations, etc. Drift is indistinguishable from a change produced by an input signal.

negative feedback Returning part of the amplified signal to the input with a phase that opposes the input signal. Although it decreases the voltage gain, it is widely used because it stabilizes the gain against transistor and temperature changes.

RF chokes Inductors that look open to RF frequencies.

split supply A power supply that has positive and negative output voltages, especially useful with differential amplifiers and operational amplifiers.

SELF-TESTING REVIEW

Read each of the following and provide the missing words. Answers appear at the beginning of the next question.

1. The most widely used method of coupling discrete circuits is _____ coupling. The coupling capacitors transmit the _____ but block the _____. The drawback in this approach is the drop in voltage gain when the frequency is too _____.

2. *(resistance-capacitance, ac, dc, low)* Occasionally, a discrete amplifier uses

Transistor Circuit Approximations

RF chokes instead of collector resistors. Such amplifiers are inherently _____ amplifiers; they are intended to amplify frequencies that are high enough for the RF chokes to appear as _____ circuits.

3. *(high-pass, open)* The one area where transformer coupling has survived is _____ amplifiers. Transformers are still used in these amplifiers because they are much smaller and less expensive than audio transformers.

4. *(radio-frequency)* With tuned transformer coupling, the collector load is resonant at one frequency and the RF amplifier provides gain over a narrow range of frequencies. At resonance, the _____ reactance cancels the capacitive reactance, leaving a purely _____ load on each collector. Tuned transformer coupling produces a bandpass amplifier.

5. *(inductive, resistive)* To extend the frequency response to zero, we can use _____ coupling. With this kind of coupling, we can get an unwanted output change known as _____. One way to reduce this unwanted change is with a differential amplifier.

6. *(direct, drift)* The cutoff frequencies of an amplifier are those frequencies where the output voltage equals 0.707 of its mid-frequency value. The lower cutoff frequency is f_1, and the upper cutoff frequency is f_2. The difference between these two frequencies is the _____. All frequencies between f_1 and f_2 are the passband of the amplifier.

7. *(bandwidth)* The alpha cutoff frequency is where the α of a transistor has dropped to 0.707 of its low-frequency value. Similarly, the _____ cutoff frequency is where the _____ of a transistor is down to 0.707 of its low-frequency value. The current-gain bandwidth product f_T is the frequency where _____ equals unity.

8. *(beta, β, β)* Hybrid parameters give exact answers, provided you have the exact h parameters for the transistor being used.

PROBLEMS

11-1. A two-stage negative-feedback amplifier has $r_F = 50$ kΩ and $r_E = 1.2$ kΩ. What is the overall voltage gain?

11-2. Suppose the RF chokes in Fig. 11-3 are 100 mH each. What is the inductive reactance of each choke at 1 MHz?

11-3. The transformer of Fig. 11-4 has a 10:1 turns ratio. If $R_L = 100$

228

Ω, what is the load resistance reflected into the primary winding? If r_e' = 40 Ω, what is the voltage gain from base to collector? The voltage gain from the base to the secondary winding?

11-4. In Fig. 11-5, $L = 200$ μH and $C = 500$ pF. What is the resonant frequency of each tuned circuit? If $r_C = 12$ kΩ, what are the Q and bandwidth of each tuned tank?

11-5. What is the dc emitter current in each transistor of Fig. 11-15a?

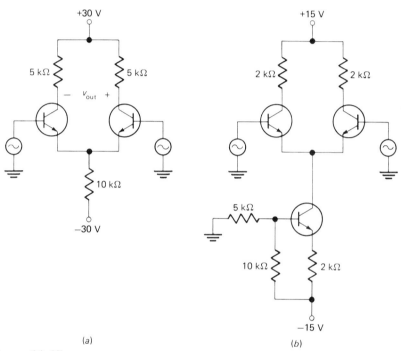

(a)   (b)

Figure 11-15

11-6. What is the voltage gain in Fig. 11-15a?

11-7. Figure 11-15b shows how a differential amplifier is biased in linear integrated circuits. Another transistor is used to source the tail current. What is the approximate value of tail current? The voltage gain?

11-8. A sine wave with a peak of 1 mV drives the diff amp of Fig. 11-16a. What is the ac output voltage?

Transistor Circuit Approximations

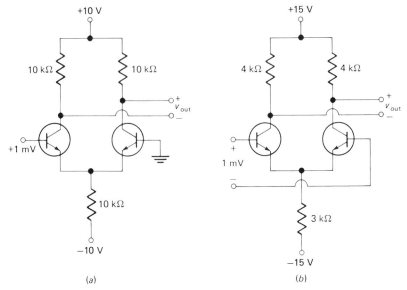

(a)   (b)

Figure 11-16

11-9. What is the voltage gain in Fig. 11-16b? The output voltage?

11-10. In Fig. 11-17, calculate the cutoff frequencies produced by the input coupling capacitor, output coupling capacitor, and emitter bypass capacitor.

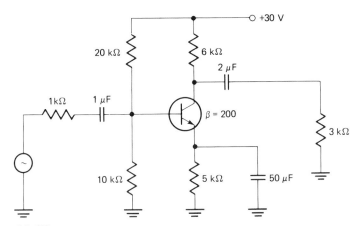

Figure 11-17

11-11. A data sheet lists an f_T of 250 MHz. What does f_α equal? If h_{fe} is 175, what does f_β equal?

230

11-12. A 2N4401 has the following minimum and maximum values for its h parameters at $I_C = 1$ mA.

Minimum: $h_{ie} = 1$ kΩ
              $h_{fe} = 40$
              $h_{re} = 0.1(10^{-4})$
              $h_{oe} = 1(10^{-6})$ μS
Maximum: $h_{ie} = 15$ kΩ
              $h_{fe} = 500$
              $h_{re} = 8(10^{-4})$
              $h_{oe} = 30$ μS

Calculate the values of α, β, r_e', and r_c' for the minimum and maximum values of h parameters.

11-13. A 2N4401 is used in a CE amplifier with an r_C of 1 kΩ. With the minimum h parameters given in Prob. 11-12, what are the values of voltage gain, current gain, input impedance, and output impedance? (Use the approximate hybrid formulas of Table 11-3.)

11-14. A CE amplifier has an R_S of 600 Ω and an r_L of 1.2 kΩ. With the exact formulas of Table 11-5 and the maximum h parameters of Prob. 11-12, calculate A, A_i, z_{in}, and z_{out}.

Chapter 12

JFETs

The bipolar transistor is the most widely used transistor in linear electronics. But there are some applications where the *junction field-effect transistor* (JFET) is preferred. This chapter examines the operation and use of JFETs.

12-1 CONSTRUCTION AND OPERATION

A bipolar transistor is called bipolar because it has both free electrons and holes as charge carriers. The JFET, however, is a *unipolar* device because it uses either free electrons or holes as majority carriers; there are no minority carriers.

JFET regions

Figure 12-1a shows parts of a JFET. The lower end is called the *source* and the upper end is the *drain.* The piece of semiconductor between the source and the drain is known as the *channel.* Since *n* material is used for the JFET in Fig. 12-1a, the majority carriers are free electrons. (For *p* material, the majority carriers would be holes.)

By embedding two *p* regions in the sides of the channel, we get the *n-channel* JFET of Fig. 12-1b. Each of these *p* regions is called a *gate.* When the manufacturer connects a separate lead to each gate, the device is called a *dual-gate* JFET. The main use of a dual-gate JFET is with a *frequency mixer,* a special circuit used in communications electronics.

This chapter concentrates on the *single-gate* JFET, a device whose gates are internally connected by the manufacturer. A single-gate JFET has only one external gate lead, as shown in Fig. 12-1c. When you see this symbol,

232

JFETs

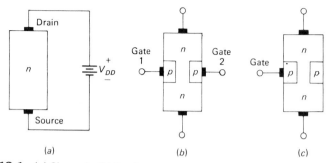

Figure 12-1 *(a)* Channel. *(b)* Dual-gate JFET. *(c)* Single-gate JFET.

remember the two *p* regions have the same potential because they are internally connected.

Biasing the JFET

Figure 12-2*a* shows the normal polarities for biasing an *n*-channel JFET. The idea is to apply a negative voltage between the gate and the source; this reverse-biases the gate. Since the gate is reverse-biased, a very small reverse current flows in the gate lead. To a first approximation, gate current is zero.

The name *field effect* is related to the depletion layers around each *pn* junction. Figure 12-2*b* shows these depletion layers. Free electrons moving between the source and the drain must flow through the narrow channel between depletion layers. The size of these depletion layers determines the width of the conducting channel. The more negative the gate voltage is, the narrower the conducting channel becomes, because the depletion layers get closer to each other. In other words, the gate voltage controls the current

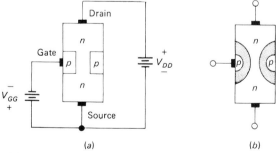

Figure 12-2 *(a)* Normal JFET bias voltages. *(b)* Depletion layers.

233

between the source and the drain. The more negative the gate voltage is, the smaller the current.

The key difference between a JFET and a bipolar transistor is this: the gate is reverse-biased, whereas the base is forward-biased. This crucial difference means the JFET is a voltage-controlled device; input voltage alone controls output current. This is different from a bipolar transistor, where input current controls output current.

We can summarize the first major difference between a JFET and a bipolar transistor in terms of resistance. The input resistance of a JFET ideally approaches infinity. To a second approximation, it is well over 10^7 Ω, depending on the particular JFET type. Therefore, in applications where high input resistance is needed, the JFET is preferred to the bipolar transistor.

The price paid for larger input resistance is smaller control over output current. In other words, a JFET is less sensitive to changes in input voltage than a bipolar transistor. In almost any JFET, an input voltage change of 0.1 V produces less than a 10-mA change in output current. But in a bipolar transistor, a 0.1-V change easily produces more than a 10-mA change in output current. As will be discussed later, this means a JFET has less voltage gain than a bipolar transistor.

Schematic symbol

Figure 12-3a shows the schematic symbol for a JFET. As a memory aid, think of the thin vertical line (Fig. 12-3b) as the channel; the source and drain connect to this line. Also, an arrow is on the gate; this arrow points to the n material. In this way, you can remember that Fig. 12-3a represents an n-channel JFET.

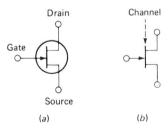

(a)   (b)

Figure 12-3 n-channel JFET symbol.

Figure 12-4 shows the p-channel JFET and its schematic symbol. The p-channel JFET is complementary to the n-channel JFET and uses holes instead of free electrons for majority carriers. Because free electrons move

234

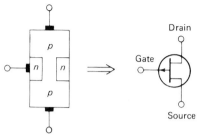

Figure 12-4 *p*-channel JFET symbol.

more easily than holes, *n*-channel devices have a better high-frequency response than *p*-channel devices.

12-2 DRAIN CURVES

Figure 12-5*a* shows a JFET with normal biasing voltages. Notice that the gate supply voltage is negative; this reverse-biases the gate and sets up the depletion layers as previously described.

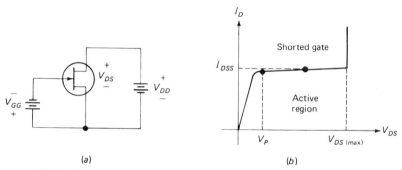

(a)   (b)

Figure 12-5 *(a)* Biased JFET. *(b)* Shorted-gate drain curve.

Pinch-off voltage

If the gate voltage is reduced to zero, the gate is effectively shorted to the source. This is called the *shorted-gate* condition. Figure 12-5*b* is a graph of drain current versus drain voltage for the shorted-gate condition. Notice the similarity to a collector curve. The drain current rises rapidly in the saturation region, but then levels off in the active region. Between voltages V_P and $V_{DS\,(max)}$, the drain current is almost constant. When the drain voltage is too large, the JFET breaks down as shown. As with a bipolar transistor,

235

the active region is along the almost horizontal part of the curve; in this region the JFET acts like a current source.

The *pinch-off voltage* is the drain voltage above which drain current becomes almost constant. When the drain voltage equals V_P, the conducting channel has become extremely narrow and the depletion layers almost touch. Because of the small passage between the depletion layers, further increases in drain voltage produce only a slight increase in drain current.

Shorted-gate drain current

In Fig. 12-5*b*, the subscripts of I_{DSS} stand for *D*rain to *S*ource with *S*horted gate. Data sheets specify I_{DSS} for an arbitrary voltage in the active region, typically between 10 and 20 V. The important thing to remember is this: because the curve is almost flat in the active region, I_{DSS} is a close approximation for the drain current anywhere in the active region for the shorted-gate condition. Furthermore, because it applies to the shorted-gate condition, I_{DSS} is the maximum drain current you can get with normal operation of the JFET; all other gate voltages are negative and result in less drain current.

Gate-source cutoff voltage

Drain curves resemble collector curves. For instance, Fig. 12-6 shows the drain curves of a typical JFET. The highest curve is for $V_{GS} = 0$, the shorted-gate condition. The pinch-off voltage is approximately 4 V, and the breakdown voltage is 30 V. As you see, I_{DSS} is 10 mA.

When $V_{GS} = -1$ V, the drain current decreases to about 5.62 mA. A V_{GS} of -2 V reduces drain current to around 2.5 mA, and so on. The bottom curve is especially important; a V_{GS} of -4 V reduces the drain current

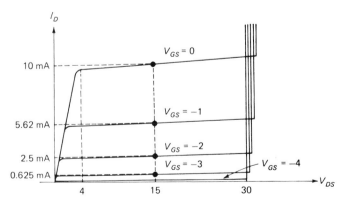

Figure 12-6 Example of drain curves.

236

to approximately zero. We call this voltage the *gate-source cutoff voltage* and symbolize it by $V_{GS(off)}$.

At $V_{GS} = V_{GS(off)}$, the depletion layers touch, cutting off the drain current. Since V_P is the drain voltage that pinches off current for the shorted-gate condition,

$$V_P = |\, V_{GS(off)} \,| \qquad\qquad \textbf{(12-1)}$$

Data sheets do not list V_P, but they give the value of $V_{GS(off)}$, which is equivalent. For instance, if you see $V_{GS(off)} = -4$ V on a data sheet, you will immediately know V_P equals 4 V.

12-3 JFET PARAMETERS

The JFET transconductance curve is a graph of I_D versus V_{GS}. By reading the value of I_D and V_{GS} in Fig. 12-6, we can plot the transconductance curve of Fig. 12-7a.

Parabolic curve

The transconductance curve of Fig. 12-7a is part of a parabola. Advanced textbooks prove the equation of this parabola is

$$I_D = I_{DSS} \left[1 - \frac{V_{GS}}{V_{GS(off)}} \right]^2 \qquad\qquad \textbf{(12-2)}$$

This is an ideal formula that can be used as an approximation for any JFET.

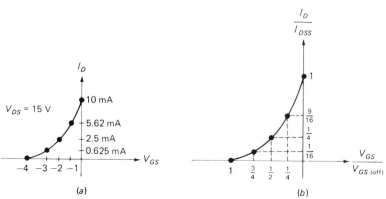

(a) (b)

Figure 12-7 Transconductance curves. *(a)* Example. *(b)* Normalized.

Square law is another name for parabolic. This is why JFETs are often called square-law devices.

Normalized transconductance curve

We can rearrange Eq. (12-2) to get

$$\frac{I_D}{I_{DSS}} = \left[1 - \frac{V_{GS}}{V_{GS\,(off)}} \right]^2 \tag{12-3}$$

By substituting 0, ¼, ½, ¾, and 1 for $V_{GS}/V_{GS\,(off)}$, we can calculate corresponding values of 1, $^9\!/_{16}$, ¼, $^1\!/_{16}$, and 0 for I_D/I_{DSS}. Figure 12-7b summarizes the results in a normalized transconductance curve. This curve applies to all JFETs.

Transconductance

The quantity g_m is called *transconductance,* defined as

$$g_m = \frac{\Delta I_D}{\Delta V_{GS}} \text{ for } V_{GS} \text{ constant} \tag{12-4}$$

This says transconductance equals the change in drain current divided by the corresponding change in gate voltage. If a change in gate voltage of 0.1 V produces a change in drain current of 0.2 mA,

$$g_m = \frac{0.2 \text{ mA}}{0.1 \text{ V}} = 2(10^{-3}) \text{ S} = 2000 \text{ } \mu\text{S}$$

Figure 12-8 brings out the meaning of g_m in terms of the transconductance curve. To calculate g_m at any quiescent point, select two nearby points that

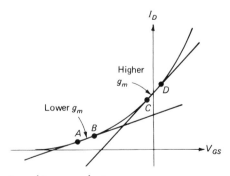

Figure 12-8 Meaning of transconductance.

straddle the quiescent point, such as A and B. The ratio of the change in I_D to the change in V_{GS} gives the value of g_m. If the Q point is further up the curve between points C and D, we get more of a change in I_D for the same change in V_{GS}; therefore, g_m is larger. In short, g_m is the *slope* of the curve at the particular operating point.

JFET data sheets almost always list the value of g_m at $V_{GS} = 0$. We will designate this particular value of g_m as g_{m0} to indicate it applies to the special case of $V_{GS} = 0$. Some data sheets also include a graph of g_m versus V_{GS}. This allows you to determine the value of g_m for any operating point. If the data sheet does not include a graph of g_m, use the following approximation:

$$g_m \cong g_{m0} \left[1 - \frac{V_{GS}}{V_{GS\,(off)}} \right] \qquad (12\text{-}5a)$$

(This equation is the derivative or slope of the transconductance curve.)

Incidentally, g_m is often designated as g_{fs} (forward transconductance) or y_{fs} (forward transadmittance). If you can't find g_m on a data sheet, look for g_{fs} or y_{fs}. As an example, the data sheet of a 2N5951 gives $g_{fs} = 6500$ μS for $V_{GS} = 0$; this means $g_{m0} = 6500$ μS.

An accurate value of $V_{GS\,(off)}$

With calculus, it's possible to derive the following formula:

$$V_{GS\,(off)} = - \frac{2I_{DSS}}{g_{m0}} \qquad (12\text{-}5b)$$

This is useful because I_{DSS} and g_{m0} are easily measured with high accuracy, but not $V_{GS\,(off)}$. Equation (12-5b) gives us a way of calculating $V_{GS\,(off)}$ with high accuracy.

Alternatively, if the pinch-off voltage V_P is known, this formula for transconductance g_{m0} may be helpful:

$$g_{m0} = \frac{2I_{DSS}}{V_P} \qquad (12\text{-}5c)$$

For example, suppose a curve tracer displays a set of JFET curves. If I_{DSS} is 10 mA and V_p is 5 V,

$$g_{m0} = \frac{2 \times 10\ \text{mA}}{5\ \text{V}} = 4000\ \mu\text{S}$$

239

12-4 SELF-BIAS

Figure 12-9*a* shows *self-bias,* the most common method for biasing a JFET. Drain current flows through R_D and R_S, producing a drain-source voltage of

$$V_{DS} = V_{DD} - I_D(R_D + R_S) \tag{12-6}$$

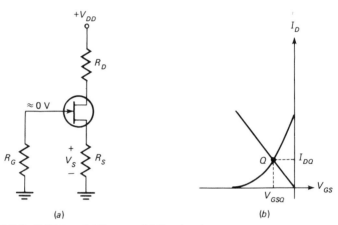

(a)   (b)

Figure 12-9 Self-bias. *(a)* Circuit. *(b)* Graphical meaning.

Because gate current is negligibly small, the gate voltage with respect to ground is approximately zero:

$$V_G = 0$$

Since drain current flows through R_S, the source voltage to ground is

$$V_S = I_D R_S$$

Therefore, the source is positive with respect to the gate, equivalent to saying the gate is negative with respect to the source. This means the gate is reverse-biased as it should be for normal operation.

Negative feedback

Self-bias is an example of negative feedback that tends to stabilize the drain current against changes in temperature and JFET replacement. For instance, suppose the drain current tries to increase in Fig. 12-9a. This produces a greater voltage drop across R_S, which means V_{GS} becomes more negative. When this happens, the depletion layers get closer together and reduce the drain current. This partially offsets the original increase in drain current.

Similarly, if the drain current tries to decrease, less voltage appears across R_S, and V_{GS} becomes less negative. This opens up the depletion layers and lets more drain current through. The overall effect is to partially compensate for the original decrease in drain current. In other words, drain current decreases but not as much as it would without the negative feedback.

The Q point

The voltage between the gate and the source is

$$V_{GS} = V_G - V_S = 0 - I_D R_S$$

or
$$V_{GS} = -I_D R_S \qquad (12\text{-}7)$$

This is a linear equation that passes through the origin when plotted as shown in Fig. 12-9b. The intersection of the source load line and the transconductance curve gives us the Q point, whose coordinates are V_{GSQ} and I_{DQ}. By rearranging Eq. (12-7), we get the following useful formula for the source resistor:

$$R_S = \frac{-V_{GSQ}}{I_{DQ}} \qquad (12\text{-}8)$$

Optimum Q point

Figure 12-10a shows the effect of different values of R_S. When R_S is too large, the Q point is too far down on the transconductance curve and the drain current is too small. On the other hand, if R_S is very small, the Q point is too near I_{DSS}. Finally, there is a correct value of R_S that sets up a Q point near the middle of the current range. In other words, ideally we'd like a quiescent drain current of $I_{DSS}/2$ to allow maximum swing in both directions. This is not a hard-and-fast rule, only a guideline.

One way to get the correct size of R_S is using Eq. (12-8) and the transconductance curve given on the data sheet of a JFET. For instance, suppose a data sheet gives a transconductance curve with an I_{DSS} of 16 mA as shown in Fig. 12-10b. Then draw a line from the origin through the 8-mA point.

241

Transistor Circuit Approximations

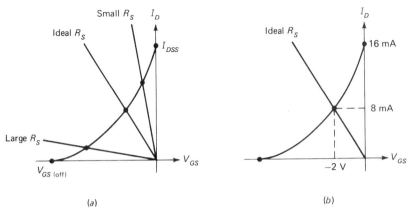

Figure 12-10 Effect of R_S on self-bias point.

Read the corresponding value of V_{GS}, which is -2 V. Then, use Eq. (12-8) to get

$$R_S = \frac{-V_{GS}}{I_D} = \frac{-(-2\,\text{V})}{8\,\text{mA}} = 250\,\Omega$$

Formula approach

If the data sheet does not show the transconductance curve, then you can calculate the optimum value of R_S using this formula:

$$R_S \cong \frac{1}{g_{m0}} \tag{12-9}$$

This equation is based on a calculus derivation that starts with Eq. (12-2). All you need to know is that the optimum R_S equals the reciprocal of g_{m0}. For instance, the data sheet of a 2N5951 lists a typical g_{m0} of 5500 μS; therefore,

$$R_S \cong \frac{1}{5500\,\mu\text{S}} = 182\,\Omega$$

Using the nearest standard size of 180 Ω in a self-biased circuit means the drain current will be approximately half of I_{DSS}.

242

Large tolerances

Some JFETs have wide tolerances in the values of I_{DSS} and g_{m0}. If the data sheet shows transconductance curves for the minimum and maximum I_{DSS}, the designer can find a compromise value of R_S as shown in Fig. 12-11. The idea is to draw a line from the origin through both curves in such a way as to get Q points somewhere near the center of both current ranges. By reading the V_{GSQ} and I_{DQ} from either curve, you can use Eq. (12-8) to calculate a compromise value of R_S.

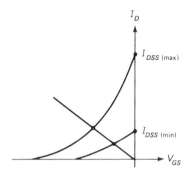

Figure 12-11 Effect of tolerance on self-bias point.

When minimum and maximum transconductance curves are not available, use the *geometric average* for g_{m0}. The formula for this average is

$$g_{m0} = \sqrt{g_{m0(min)}g_{m0(max)}} \qquad (12\text{-}10)$$

As an example, the data sheet of a 2N5457 lists a $g_{m0(min)}$ of 2000 μS and a $g_{m0(max)}$ of 7000 μS. The geometric average is

$$g_{m0} = \sqrt{2000\ \mu S \times 7000\ \mu S} = 3742\ \mu S$$

With Eq. (12-9),

$$R_S \cong \frac{1}{g_{m0}} = \frac{1}{3742\ \mu S} = 267\ \Omega$$

Using the nearest standard size of 270 Ω gives a compromise value of R_S that works well for all 2N5457s.

243

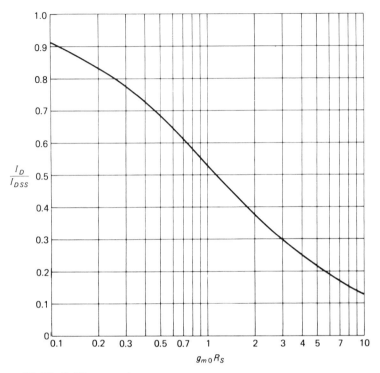

Figure 12-12 Self-bias graph.

The universal self-bias graph

Starting with Eqs. (12-2) and (12-7), it is possible with calculus to derive the relationships between drain current, shorted-gate drain current, transconductance, and source resistance. Figure 12-12 shows a graph of these quantities and allows us to analyze any self-biased circuit. When the $g_{m0}R_S$ product decreases, the I_D/I_{DSS} ratio increases. Conversely, when $g_{m0}R_S$ increases, I_D/I_{DSS} decreases. Notice also that when $g_{m0}R_S = 1$, $I_D/I_{DSS} = 0.53$. This means drain current is approximately half of I_{DSS} when

$$g_{m0}R_S = 1$$

or

$$R_S = \frac{1}{g_{m0}}$$

This agrees with the earlier discussion; resistance R_S should equal the reciprocal of g_{m0} to get an I_D approximately equal to half of I_{DSS}.

244

EXAMPLE 12-1

The JFET of Fig. 12-13a has a g_{mo} of 3000 µS and an I_{DSS} of 2 mA. What is the dc voltage from the drain to ground? From the source to ground?

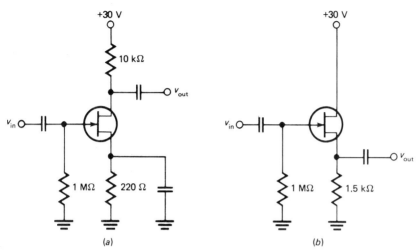

Figure 12-13

SOLUTION

The capacitors have nothing to do with the problem because they appear open to dc. First, get the $g_{mo}R_S$ product:

$$g_{mo}R_S = 3000 \text{ µS} \times 220 \text{ } \Omega = 0.66$$

With Fig. 12-12, we can read the approximate value of the current ratio:

$$\frac{I_D}{I_{DSS}} = 0.62$$

or $$I_D = 0.62 I_{DSS} = 0.62 \times 2 \text{ mA} = 1.24 \text{ mA}$$

The dc drain-ground voltage equals the supply voltage minus the drop across the drain resistor:

$$V_D = V_{DD} - I_D R_D = 30 \text{ V} - (1.24 \text{ mA} \times 10 \text{ k}\Omega) = 17.6 \text{ V}$$

The dc source-ground voltage equals the drain current times the source resistance:

245

$$V_S = I_D R_S = 1.24 \text{ mA} \times 220 \text{ }\Omega = 0.273 \text{ V}$$

EXAMPLE 12-2

The JFET in Fig. 12-13b has the same specifications as in the preceding example: $g_{m0} = 3000$ μS and $I_{DSS} = 2$ mA. What are the dc drain current, and the dc voltage between the source and ground?

SOLUTION

As usual, start with the $g_{m0} R_S$ product:

$$g_{m0} R_S = 3000 \text{ }\mu\text{S} \times 1.5 \text{ k}\Omega = 4.5$$

With Fig. 12-12, read the corresponding current ratio, which is approximately

$$\frac{I_D}{I_{DSS}} = 0.23$$

This gives a drain current of

$$I_D = 0.23 I_{DSS} = 0.23 \times 2 \text{ mA} = 0.46 \text{ mA}$$

and a source-to-ground voltage of

$$V_S = I_D R_S = 0.46 \text{ mA} \times 1.5 \text{ k}\Omega = 0.69 \text{ V}$$

12-5 CURRENT-SOURCE BIAS

As mentioned earlier, self-bias uses negative feedback to stabilize drain current against changes in temperature and JFET replacement. But the compensation is not perfect. This section discusses *current-source* bias, the ultimate weapon for holding drain current constant in spite of large changes in JFET parameters.

Two supplies

Figure 12-14a shows how it's done when a split supply is available. The bipolar transistor acts like a current source and forces the JFET current to equal the collector current. Since the bipolar transistor uses emitter bias, the approximate emitter current is

$$I_E \cong \frac{V_{EE}}{R_E} \tag{12-11}$$

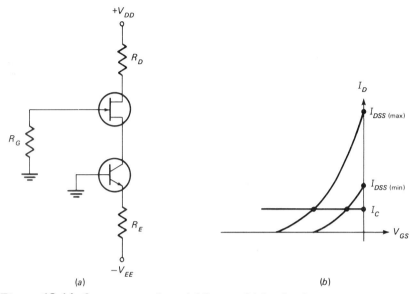

Figure 12-14 Current-source bias. *(a)* Circuit. *(b)* Graphical meaning.

The collector diode acts like a current source and forces I_D to equal I_C, which in turn approximately equals I_E.

With current-source bias, the collector current must always be less than the shorted-gate drain current; otherwise, the gate is driven into forward bias. In other words, we must satisfy this condition:

$$I_C < I_{DSS} \qquad \textbf{(12-12)}$$

This guarantees that V_{GS} is negative.

Figure 12-14*b* brings out the importance of condition (12-12). Here we have the minimum and maximum transconductance curves of a JFET. The bipolar transistor sets up a constant current of I_C. As long as I_C is less than $I_{DSS\,(min)}$, operation is normal in spite of the JFET tolerances. For instance, a 2N5952 has $I_{DSS\,(min)} = 4$ mA and $I_{DSS\,(max)} = 8$ mA. To work with all 2N5952s, current-source bias must set up a collector current of less than 4 mA.

One supply

When you do not have a negative supply, you can still use current-source bias as shown in Fig. 12-15. In this circuit, we have voltage-divider bias on the bipolar transistor. Almost all the voltage across R_2 appears across

247

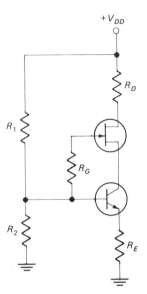

Figure 12-15 Single-supply current-source bias.

the R_E resistor. This fixes the emitter current to a constant value, forcing the drain current to be independent of JFET variations.

12-6 JFET AMPLIFIERS

The complete analysis of JFET amplifiers is too complicated to reproduce here. What we will do in this section is discuss the three basic configurations, stating some gain formulas without proof. (For an in-depth discussion, see Albert P. Malvino, *Electronic Principles,* Second Edition, McGraw-Hill Book Company, New York, 1979, pages 358–366, 430–437.)

Common-source amplifier

Figure 12-16 is a common-source amplifier. The bypass capacitor places the source terminal at ac ground. Because of the output coupling capacitor, the drain and load resistors appear in parallel. In other words, the ac resistance seen by the drain is

$$r_D = R_D \parallel R_L$$

Because the source terminal is at ac ground, the ac input signal appears across the gate-source terminals, and the change in gate-source voltage is

$$\Delta V_{GS} = v_{\text{in}}$$

248

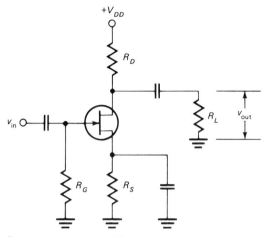

Figure 12-16 Common-source amplifier.

Since $g_m = \Delta I_D/\Delta V_{GS}$, the change in drain current produced by the ac input signal is

$$\Delta I_D = g_m \Delta V_{GS} = g_m v_{\text{in}}$$

When this ac drain current flows through the ac drain resistance, we get an ac output voltage of

$$v_{\text{out}} = \Delta I_D r_D = g_m v_{\text{in}} r_D$$

or

$$\frac{v_{\text{out}}}{v_{\text{in}}} = g_m r_D$$

Therefore, the voltage-gain formula is

$$A = g_m r_D \qquad\qquad \textbf{(12-13)}$$

So, if $g_m = 4000\ \mu\text{S}$ and $r_D = 5\ \text{k}\Omega$,

$$A = 4000\ \mu\text{S} \times 5\ \text{k}\Omega = 20$$

As mentioned earlier, the input resistance of a JFET is extremely large because the gate is reverse-biased. Typically, this input resistance is well into the megohms, so that the input impedance of the stage at low frequencies is approximately equal to the gate-return resistor:

$$z_{\text{in}} \cong R_G \qquad\qquad \textbf{(12-14)}$$

249

Transistor Circuit Approximations

If $R_G = 1$ MΩ, the stage has an input impedance of approximately 1 MΩ at low frequencies.

Sometimes, part of the source resistor is left unbypassed (similar to a swamping resistor in bipolar amplifiers). In this case, the formula for the voltage gain is

$$A = \frac{g_m r_D}{1 + g_m r_S} \qquad (12\text{-}15)$$

where r_S is the swamping resistor.

Getting the transconductance

Many data sheets include a graph of g_m versus drain current. This allows you to determine the value of g_m at the quiescent point. Then, with Eq. (12-13) or (12-15) you can calculate the voltage gain for a common-source amplifier. If the data sheet does not include a graph of g_m versus drain current, use Fig. 12-17 to work out the g_m at the Q point. For instance, if $I_D/I_{DSS} = 0.5$, the corresponding g_m in Fig. 12-17 is

$$\frac{g_m}{g_{m0}} = 0.7$$

or
$$g_m = 0.7 g_{m0} \qquad (12\text{-}16)$$

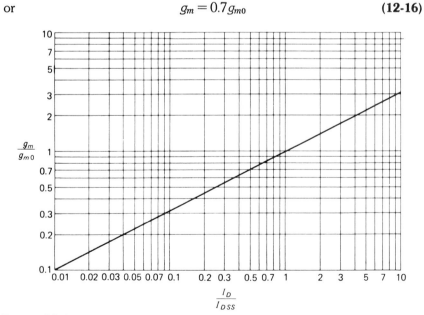

Figure 12-17 Variation of g_m with operating point.

250

This says the g_m equals 70 percent of g_{mo}, the value listed on almost all data sheets.

Common-drain amplifier

Figure 12-18 shows the common-drain amplifier, better known as the *source follower.* This is analogous to an emitter follower. The voltage gain approaches unity, and the output signal is in phase with the input signal. The whole purpose of the source follower is to present a high input impedance to the ac input signal. With a source follower, this input impedance can

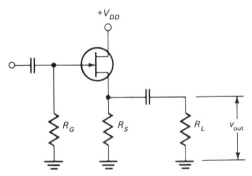

Figure 12-18 Source follower.

be in the megohms, depending on the size of R_G. As a result, the source follower has even higher input impedance than the emitter follower, and this is why source followers are used a lot at the front end of measuring instruments like voltmeters and oscilloscopes.

It can be shown that the voltage gain of a source follower is

$$A = \frac{g_m r_s}{1 + g_m r_s} \qquad \textbf{(12-17)}$$

where $r_s = R_S \| R_L$, the ac resistance seen by the source terminal. Again, the input impedance at low frequencies is approximately equal to the gate-return resistor:

$$z_{\text{in}} \cong R_G$$

Common-gate amplifier

The applications of a common-gate amplifier (Fig. 12-19) are limited. It has been used at high frequencies and in *cascode* amplifiers (a common-

251

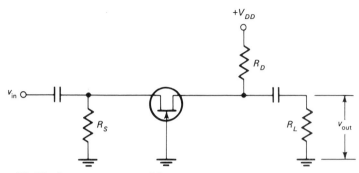

Figure 12-19 Common-gate amplifier.

source driving a common-gate amplifier). Aside from these special applications, the common-gate amplifier is rarely used. One reason is its low input
impedance. Notice the input signal has to drive the source terminal; therefore,
the ac input current equals the ac drain current. As shown earlier, $\Delta I_D = g_m v_{\text{in}}$. As a result, the ac input current is

$$i_{\text{in}} = \Delta I_D = g_m v_{\text{in}}$$

or

$$\frac{v_{\text{in}}}{i_{\text{in}}} = \frac{1}{g_m}$$

This means the input impedance is relatively low, equal to

$$z_{\text{in}} = \frac{1}{g_m} \tag{12-18}$$

With an analysis similar to that given for the common-source amplifier,
the voltage gain of a common-gate amplifier is

$$A = g_m r_D$$

12-7 THE JFET USED AS A SWITCH

One of the major applications of a JFET is switching. The idea is to use
only two points on the load line: cutoff and saturation. When the JFET is
cut off, it is like an open switch; when it's saturated, it's like a closed switch.

Basic idea

Figure 12-20 shows the shorted-gate drain curve. When $V_{GS} = 0$, the operating
point is at saturation and the JFET is approximately equivalent to a closed

252

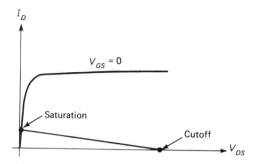

Figure 12-20 Switching operation of JFET.

switch. To open the switch, we have to apply a gate voltage more negative than $V_{GS\,(off)}$; this drives the JFET into cutoff. The region between saturation and cutoff is not used in switching operation. In other words, only two points are involved on the load line; we get this two-state operation by applying either a low gate voltage (zero) or a high gate voltage (greater than the gate-source cutoff voltage).

DC on-state resistance

The switching action is not perfect because the JFET has a small resistance when it's saturated. The *static* or *dc on-state resistance* is the ratio of drain voltage to drain current. As a formula,

$$r_{DS\,(on)} = \frac{V_{DS}}{I_D} \qquad (12\text{-}19)$$

As an example, suppose the saturation point of Fig. 12-20 has $V_{DS} = 0.1$ V and $I_D = 0.8$ mA. Then, the dc on-state resistance is

$$r_{DS\,(on)} = \frac{0.1 \text{ V}}{0.8 \text{ mA}} = 125 \text{ } \Omega$$

AC on-state resistance

When used as a switch, the JFET is normally saturated well below the knee of the drain curve as shown in Fig. 12-20. In other words, the drain current is much smaller than I_{DSS} and the drain voltage is much smaller than V_p. If small ac signals pass through the JFET when it's saturated, it presents an ac resistance to these signals, defined as

253

Transistor Circuit Approximations

$$r_{ds\,(\text{on})} = \frac{\Delta V_{DS}}{\Delta I_D} \qquad\qquad \textbf{(12-20)}$$

This ac resistance, also known as the *small-signal on-state* resistance, equals the slope of the drain curve at the saturation point.

As mentioned, the saturation point is almost always well below the knee where the drain curve is linear. For this reason, $r_{ds\,(\text{on})}$ and $r_{DS\,(\text{on})}$ are approximately equal. In other words, when the JFET acts like a closed switch, it has the same resistance to ac and dc signals.

Shunt switching

Figure 12-21a is a *shunt switch*. The idea is to either transmit or block the input signal v_{in}. When $V_{\text{CONTROL}} = 0$, the JFET is equivalent to a closed switch and $v_{\text{out}} = 0$. When V_{CONTROL} is more negative than $V_{GS\,(\text{off})}$, the JFET is like an open switch and v_{out} equals v_{in}.

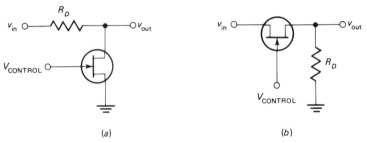

Figure 12-21 JFET switching circuits. *(a)* Shunt. *(b)* Series.

The switching is not perfect because of $r_{ds\,(\text{on})}$. Resistance $r_{ds\,(\text{on})}$ and R_D form a voltage divider, so that we do get a small output signal even when the JFET acts like a closed switch. A designer tries to make R_D much greater than $r_{ds\,(\text{on})}$ to minimize the output signal when V_{CONTROL} is zero. For instance, a 2N4391 has an $r_{ds\,(\text{on})}$ of 30 Ω; if R_D is 30 kΩ, only 1/1000 of the input signal reaches the output when the JFET is closed.

Series switching

Figure 12-21b is a *series switch*. Again, the point is to transmit or block the input signal. When V_{CONTROL} is zero, the JFET is closed and the output equals the input. But when V_{CONTROL} is more negative than $V_{GS\,(\text{off})}$, the JFET is open, and the output signal approaches zero. As before, $r_{ds\,(\text{on})}$ should be much smaller than R_D.

254

Analog switches

The series and shunt switches just described are examples of *analog switches*, those intended to transmit or block ac signals. These signals can be as elementary as sine waves or as complex as voice and music. The applications for analog switching are so widespread that manufacturers are producing analog switching ICs; some of these chips include JFET switches with bipolar drivers.

Multiplexing

Multiplex means "many into one." Figure 12-22*a* shows a 3-to-1 multiplexer, a circuit that steers one of the input signals to the output line. When the control signals (V_1, V_2, and V_3) are more negative than $V_{GS\,(off)}$, all input

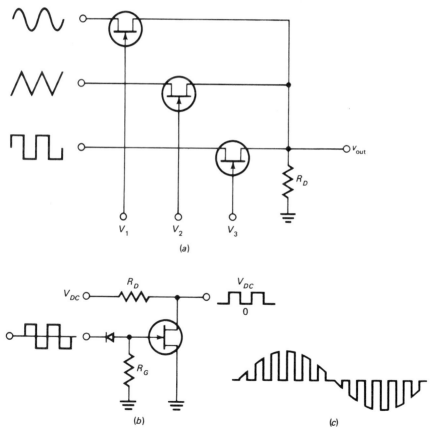

Figure 12-22 *(a)* Three-to-one multiplexer. *(b)* JFET chopper.

255

signals are blocked. By making any control signal equal to zero, we can transmit one of the inputs to the output. For instance, if $V_1 = 0$, we get a sinusoidal output. If $V_2 = 0$, we get a triangular output. And if $V_3 = 0$, we get a square-wave output. Normally, only one of the control signals equals zero.

IC multiplexers are available that can handle from 2 to 16 inputs.

Choppers

As discussed earlier, we can build a direct-coupled amplifier by leaving out the coupling and bypass capacitors, and connecting the output of each stage directly to the input of the next stage. In this way, dc is coupled, as well as ac. But the major disadvantage of direct coupling is drift, a slow shift in the final output voltage produced by supply and transistor variations.

Figure 12-22b shows a JFET *chopper,* one way of getting around the drift problem. Here is the idea. The input signal to be amplified is V_{DC}. By applying a large square wave to the gate, we can force the JFET to close and open; the diode on the gate lead prevents the positive half cycle from forward-biasing the gate. When the JFET is closed, the output is ideally zero. When the JFET is open, the output is ideally V_{DC}. Since the JFET is alternately closing and opening, we get the chopped output shown in Fig. 12-22b. This output swings from 0 to V_{DC}.

The chopped output can now go to a conventional ac amplifier, one with coupling capacitors between stages; this eliminates the drift problem. The amplified square-wave output can then be peak-detected to recover the dc signal.

The idea also works with low-frequency ac signals. For instance, if a sine wave replaces V_{DC}, the chopped output looks like Fig. 12-22c. This signal then goes to an ac amplifier and peak detector. The only restriction is that the chopper frequency (gate signal) must be much higher than the input signal being amplified.

GLOSSARY

cascode With JFETs this means a common-source driving a common-gate connection. With bipolar transistors it's a common-emitter driving a common-base connection.

chopper A device or circuit that can chop a dc input signal to get an ac output signal.

frequency mixer A communications circuit whose output frequency equals the difference of two input frequencies.

256

multiplexer A circuit with many inputs but only one output. Control signals allow you to steer one of the input signals to the output line.

on-state resistance The resistance between the drain and source terminals of a JFET being operated as a closed switch.

pinch-off voltage On the shorted-gate drain curve, the pinch-off voltage is the approximate drain voltage above which drain current becomes essentially constant.

self-bias The most common type of JFET biasing. Drain current through a source resistor produces a reverse bias on the gate-source terminals.

source follower A widely used JFET input circuit. It has the advantage of extremely high input resistance.

transconductance The ratio of a change in drain current to a change in gate-source voltage, measured under the condition of a constant drain-source voltage.

SELF-TESTING REVIEW

Read each of the following and provide the missing words. Answers appear at the beginning of the next question.

1. The name *field effect* is related to the _____ layers of a JFET. The more negative the gate voltage, the narrower the conducting _____ becomes.

2. *(depletion, channel)* The input resistance of a JFET ideally approaches _____. To a second approximation, it is well into the megohms. In applications where _____ input resistance is needed, the JFET is preferred to the bipolar transistor.

3. *(infinity, high)* The pinch-off voltage is the drain voltage on the shorted-gate drain curve above which the drain current becomes almost _____. The drain current along this almost constant part of the curve is designated _____.

4. *(constant, I_{DSS})* The gate-source voltage that reduces the drain current to approximately zero is called the gate-source _____ voltage, symbolized $V_{GS(off)}$. The absolute values of pinch-off voltage and gate-source cutoff voltage are _____.

5. *(cutoff, equal)* Transconductance g_m is also symbolized as _____ or _____. The g_m at $V_{GS} = 0$ is designated as g_{m0}. Almost all data sheets list g_{m0}.

Transistor Circuit Approximations

6. *(g_{fs}, y_{fs})* The most common method for biasing a JFET is _____ bias. This type of bias is an example of negative _____ that tends to stabilize drain current against changes in temperature and JFET replacement.

7. *(self, feedback)* Current-source bias is the ultimate weapon for holding drain current _____ in spite of large changes in JFET parameters. With this kind of bias, we must keep collector current less than _____.

8. *(constant, I_{DSS})* The common-drain amplifier, better known as the _____ follower, has a voltage gain approaching unity and the output signal is in _____ with the input signal. This circuit is used a lot at the front end of measuring instruments like voltmeters and oscilloscopes.

9. *(source, phase)* One of the major applications of a JFET is _____. The idea is to use only _____ points on the load line: cutoff and saturation. When a JFET is cut off, it's like an open switch. When it's saturated, it's like a closed switch.

10. *(switching, two)* Some JFET applications are analog switching, multiplexing, and chopping.

PROBLEMS

12-1. At room temperature a 2N4220 has a reverse gate current of 0.1 nA for a reverse gate voltage of 15 V. Calculate the resistance looking into the gate.

12-2. A JFET has a $V_{GS(off)}$ of −3 V. What is its pinch-off voltage?

12-3. I_{DSS} = 12 mA and $V_{GS(off)}$ = −5 V. Use Eq. (12-2) to calculate the drain current for a gate-source voltage of −2 V.

12-4. When V_{DS} = 10 V, a change of 0.2 V in V_{GS} produces a change of 0.65 mA in I_D. What does the transconductance equal?

12-5. The g_{m0} of a JFET equals 5000 μS. If $V_{GS(off)}$ is −4 V, what does g_m equal when V_{GS} is −2 V?

12-6. In a self-biased circuit, a drain current of 3 mA flows through a source resistor of 1.5 kΩ. What does V_{GS} equal?

12-7. A JFET has a g_{m0} of 4000 μS. To bias this JFET near the middle of its current range, what value of R_S should we use in a self-biased circuit?

258

12-8. The data sheet of a JFET gives minimum and maximum g_{mo} values of 2500 μS and 7500 μS. What is the geometric average of g_{mo}? If the JFET is to be used in a self-biased circuit, what value of R_S gives a good compromise for operating near the middle of the current range?

12-9. The JFET of Fig. 12-23a has g_{mo} = 5000 μS and I_{DSS} = 15 mA. What is the dc voltage from the drain to ground? From the source to ground?

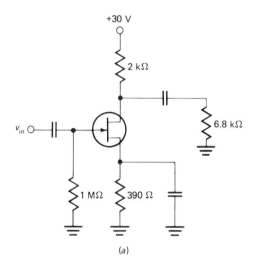

(a)

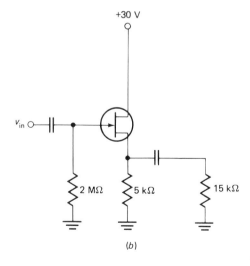

(b)

Figure 12-23

Transistor Circuit Approximations

12-10. In Fig. 12-23b, g_{mo} = 2000 μS and I_{DSS} = 12 mA. What is the dc voltage from source to ground?

12-11. In Fig. 12-14a, V_{EE} = 30 V and R_E = 10 kΩ. What is the approximate value of drain current? If R_D = 5 kΩ and V_{DD} = 30 V, what is the dc voltage from the drain to ground? (Note: I_{DSS} = 10 mA.)

12-12. If the JFET of Fig. 12-23a has a g_m of 3000 μS at the Q point, what is the voltage gain? The input resistance of the stage?

12-13. The JFET in Fig. 12-23b has g_m = 1500 μS at the quiescent point. What is the voltage gain of the source follower? Its input resistance?

12-14. A common-gate amplifier has a g_m of 4000 μS at its Q point. What is the input resistance looking into the source?

12-15. A 2N5114 has an $r_{DS(on)}$ of 75 Ω. If 1 mA flows through the JFET when it's used as a closed switch, what is the voltage between the drain and the source?

12-16. The data sheet of a 2N3684 lists an $r_{DS(on)}$ of 600 Ω. What does $r_{ds(on)}$ equal if this JFET is operating well below the pinch-off voltage?

12-17. The 2N3972 of Fig. 12-24a has an $r_{DS(on)}$ of 100 Ω and a $V_{GS(off)}$ of −3 V. If v_{in} = 8 V, what does v_{out} equal when V_{GS} is 0 V? When V_{GS} = −3 V?

(a)   (b)

Figure 12-24

12-18. In Fig. 12-24b, the 2N3966 has an $r_{DS(on)}$ of 220 Ω and a $V_{GS(off)}$ of −6 V. What does v_{out} equal if v_{in} is 10 V and V_{GS} is 0 V? If V_{GS} = −6 V, what does v_{out} equal?

260

Chapter 13

MOSFETs

Computers use integrated circuits with thousands of transistors. These integrated circuits work remarkably well, despite the effects of changing temperature and transistor tolerance. How is it possible? The answer is *two-state* design, using only two points on the load line of each transistor. In other words, each transistor acts like a *switch* rather than a linear amplifier.

This is where the metal-oxide semiconductor field-effect transistor (MOSFET) has made its greatest impact. It is almost ideal as a switching device because of its low power consumption. Furthermore, MOS circuits take up much less chip area than bipolar circuits. This is an important advantage in *large-scale integration* (LSI), chips with incredibly complex circuits.

13-1 DEPLETION-TYPE MOSFET

Like the JFET, the MOSFET has a source, gate, and drain. The big difference is that the gate is insulated from the channel. Because of this, we can apply positive voltages as well as negative voltages to the gate. In either case, negligible gate current exists.

MOSFET regions

To begin with, there's an *n* region with a source and drain as shown in Fig. 13-1*a*. As before, a positive voltage applied to the drain-source terminals forces free electrons to flow from the source to the drain. Unlike the JFET, the MOSFET has only a single *p* region as shown in Fig. 13-1*b*. This region is called the *substrate*. This *p* region constricts the channel between source and drain so that only a small passage remains at the left side of Fig. 13-

261

Transistor Circuit Approximations

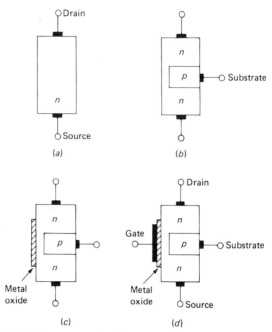

Figure 13-1 MOSFET structure. *(a)* The *n* channel. *(b)* Adding the substrate. *(c)* Adding the metal oxide. *(d)* Adding the gate.

1*b*. Electrons flowing from source to drain must pass through this narrow channel.

A thin layer of metal oxide (usually silicon dioxide) is deposited over the left side of the channel as shown in Fig. 13-1*c*. This metal oxide is an insulator. Finally, a metallic gate is deposited on the insulator (Fig. 13-1*d*). Because the gate is insulated from the channel, a MOSFET is also known as an insulated-gate FET (IGFET).

Depletion mode of operation

How does the MOSFET of Fig. 13-2*a* work? As usual, the V_{DD} supply forces free electrons to flow from source to drain. These electrons flow through the narrow channel to the left of the *p* substrate. As before, the gate voltage controls the resistance of the *n* channel. But since the gate is insulated from the channel, we can apply either a positive or negative voltage to the gate. Figure 13-2*a* shows a negative gate voltage. This repels free electrons in the channel, pushing them back toward the source. In other

262

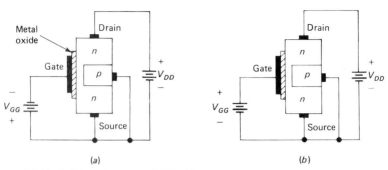

Figure 13-2 *(a)* Negative gate. *(b)* Positive gate.

words, a negative gate voltage reduces the flow between the source and the drain.

The more negative the gate voltage is, the greater the depletion of charges in the channel. Enough negative gate voltage cuts off the current between the source and the drain. Therefore, with negative gate voltage the action of a MOSFET is similar to that of a JFET. Because the action depends on reducing or *depleting* the charges in the channel, negative gate operation is known as the *depletion mode*.

Enhancement mode

Figure 13-2*b* shows a positive voltage applied to the gate. This attracts free electrons into the channel and increases the current between the source and the drain. In other words, a positive gate voltage *enhances* the conductivity of the channel. The more positive the gate voltage, the greater the conduction from source to drain. Operating the MOSFET with a positive gate voltage is called the *enhancement mode*.

Because of the insulating layer, negligible gate current flows in either mode of operation. In fact, the input resistance of a MOSFET is incredibly high, typically from 10^{10} to over 10^{14} Ω.

The device in Fig. 13-2 is an *n*-channel MOSFET; the complementary device is a *p*-channel MOSFET. We will concentrate on *n*-channel devices and extend the results to *p*-channel ones at a later time.

MOSFET curves

Figure 13-3*a* shows typical drain curves for an *n*-channel MOSFET. $V_{GS(\text{off})}$ represents the negative gate voltage that cuts off drain current. For V_{GS}

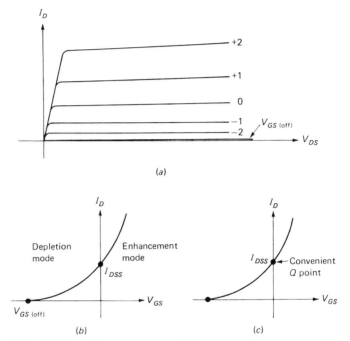

Figure 13-3 *(a)* MOSFET drain curves. *(b)* Transconductance curve. *(c)* Zero bias.

less than zero, we get depletion-mode operation. On the other hand, V_{GS} greater than zero gives enhancement-mode operation.

Figure 13-3*b* is the transconductance curve. I_{DSS} is the drain-source current with a shorted gate. I_{DSS} no longer is the maximum possible current; the curve now extends to the right of the origin. MOSFETs with a transconductance curve like Fig. 13-3*c* are easier to bias than JFETs because they can use a Q point with zero gate voltage as shown.

Any MOSFET that can operate in either the depletion or enhancement mode is called a *depletion-type* MOSFET. Since this type of MOSFET has drain current with zero gate voltage, it is also called a *normally on* MOSFET.

Schematic symbol

Figure 13-4*a* shows the schematic symbol for a normally on MOSFET. The gate appears like a capacitor plate. Just to the right of the gate is a thin vertical line representing the channel; the drain lead comes out the top of the channel and the source lead connects to the bottom. The arrow on

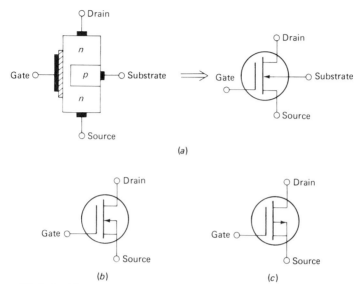

Figure 13-4 D-MOSFET symbols. *(a)* n channel with substrate lead. *(b)* n channel device. *(c)* p-channel device.

the substrate points to the *n* material; therefore, the device is an *n*-channel MOSFET.

Usually, the manufacturer internally connects the substrate to the source; this results in a three-terminal device whose schematic symbol is shown in Fig. 13-4*b*. By using the opposite type of doping, a manufacturer can produce a *p*-channel MOSFET, whose schematic symbol is shown in Fig. 13-4*c*.

13-2 ENHANCEMENT-TYPE MOSFET

There is another kind of MOSFET, the *enhancement type,* that is much more important when it comes to digital circuits. It can operate only in the enhancement mode.

Creating the inversion layer

Figure 13-5*a* shows the different parts of an enhancement-type MOSFET. Notice the substrate extends all the way to the metal oxide; there no longer is an *n* channel between the source and the drain.

How does the enhancement-type MOSFET conduct? Figure 13-5*b* shows the normal biasing polarities. When $V_{GS} = 0$, the V_{DD} supply tries to force free electrons to flow from the source to the drain, but the *p* substrate has

265

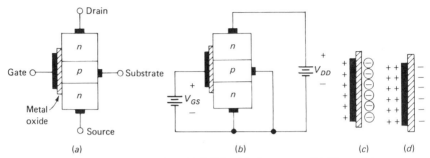

Figure 13-5 Enhancement-only MOSFET. *(a)* Structure. *(b)* Normal bias. *(c)* Creation of negative ions. *(d)* Creation of *n*-type inversion layer.

only a few thermally produced free electrons. Aside from these minority carriers and some surface leakage, the current between source and drain is zero. For this reason, the enhancement-type MOSFET is a *normally off* MOSFET.

To get drain current, we have to apply enough positive voltage to the gate. The gate acts like one plate of a capacitor, the metal oxide like a dielectric, and the *p* substrate like the other plate. For smaller gate voltages the positive charges in Fig. 13-5*c* induces negative charges in the *p* substrate. These charges are negative *ions,* produced by valence electrons filling holes in the *p* substrate. With a further increase in gate voltage, the additional positive charges on the gate can put free electrons into orbit around the negative ions (see Fig. 13-5*d*). In other words, when the gate is positive enough, it can create a thin layer of free electrons stretching all the way from the source to the drain.

The created layer of free electrons is next to the metal oxide. This layer no longer acts like a *p*-type semiconductor. Instead, it appears like an *n*-type semiconductor because of the induced free electrons. For this reason, the layer of *p* material touching the metal oxide is called an *n-type inversion layer.*

The threshold voltage

The minimum gate-source voltage that creates the *n*-type inversion layer is called the *threshold voltage,* designated $V_{GS(th)}$. When the gate voltage is less than the threshold voltage, zero current flows from source to drain. But when the gate voltage is greater than the threshold voltage, an *n*-type inversion layer connects the source to the drain and we get current.

Threshold voltages depend on the particular type of MOSFET. $V_{GS(th)}$ can vary from less than a volt to more than 5 V. For instance, the 3N169 is an enhancement-type MOSFET; it has a threshold voltage of 1.5 V.

266

The enhancement-type MOSFET is a natural choice for digital circuits because this type of MOSFET is normally off. When the gate voltage exceeds the threshold voltage, the MOSFET turns on like a switch. Later sections discuss some digital circuits using enhancement-type MOSFETs.

Enhancement-type curves

Figure 13-6 shows a set of curves for an enhancement-type MOSFET. The lowest curve is the $V_{GS(th)}$ curve. When V_{GS} is less than $V_{GS(th)}$, the drain current is ideally zero and the MOSFET is off. When V_{GS} is greater than $V_{GS(th)}$, drain current appears. The larger V_{GS} is, the greater the drain current.

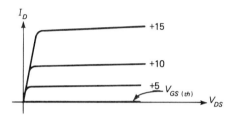

Figure 13-6 Drain curves for E MOSFET.

Schematic symbols

When $V_{GS} = 0$, the enhancement-type MOSFET is off, because no conducting channel exists between the source and the drain. The schematic symbol of Fig. 13-7a has a broken channel line to indicate this normally off condition. As we know, a gate voltage greater than the threshold voltage creates an n-type inversion layer that connects the source to the drain. The arrow points to this inversion layer, which acts like an n channel when the device is conducting. For this reason, the device is an n-channel enhancement-type MOSFET.

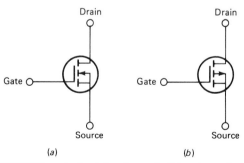

Figure 13-7 E-MOSFET symbols. *(a)* n channel. *(b)* p channel.

267

Figure 13-7b shows the schematic symbol for the complementary MOSFET, a p-channel enhancement type. In this case, the threshold voltages are negative and drain current is in the opposite direction from an n-channel device.

Maximum gate-source voltage

Both depletion-type and enhancement-type MOSFETs have a thin insulating layer between the gate and the channel. This thin layer is easily destroyed by excessive gate-source voltage. For instance, a 2N3796 has a $V_{GS(max)}$ rating of ±30 V. If the gate-source voltage becomes more positive than +30 V or more negative than −30 V, you can throw away the MOSFET because the thin insulating layer has been destroyed.

Aside from directly applying an excessive V_{GS}, you can destroy the thin insulating layer in more subtle ways. Remove or insert a MOSFET into a circuit while the power is on, and transient voltages may exceed $V_{GS(max)}$, ruining the MOSFET. Even picking up a MOSFET may deposit enough static charge to exceed the $V_{GS(max)}$ rating. This is the reason MOSFETs are often shipped with a wire ring around the leads. The ring is removed after the MOSFET is connected in a circuit.

The newer MOSFETs are protected by built-in zener diodes in parallel with the gate and the source. The zener voltage is less than the $V_{GS(max)}$ rating. In this way, the zener diode breaks down before any damage occurs to the thin insulating layer. The disadvantage of these built-in zener diodes is that they reduce the MOSFET's high input resistance.

13-3 BIASING MOSFETS

With depletion-type MOSFETs (D MOSFETs), V_{GS} can be positive, negative, or zero. But with enhancement-type MOSFETs (E MOSFETs), V_{GS} has to be greater than the threshold voltage to get drain current.

Zero bias of D MOSFETs

Since a D MOSFET can operate in either the depletion or enhancement mode, we can set its Q point at $V_{GS} = 0$ as shown in Fig. 13-8a. Then, an ac input signal to the gate can produce variations above and below the Q point. Being able to use zero V_{GS} is an advantage because it permits us to use the unique biasing circuit of Fig. 13-8b. This simple circuit has $V_{GS} = 0$ and $I_D = I_{DSS}$. This *zero bias* is unique with D MOSFETs; it will not work with a bipolar transistor, JFET, or E MOSFET.

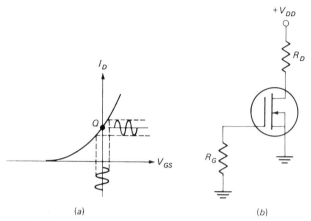

(a)   (b)

Figure 13-8 Zero bias.

Drain-feedback bias of E MOSFETs

Figure 13-9a shows *drain-feedback bias,* a type of bias used with E MOSFETs. Since gate current is negligible, no voltage appears across R_G; therefore, $V_{GS} = V_{DS}$. Like collector-feedback bias, the circuit of Fig. 13-9a tends to compensate for changes in MOSFET characteristics. If I_D tries to increase because of temperature change or MOSFET replacement, drain-source voltage V_{DS} will decrease. This reduces the gate bias voltage V_{GS}, which partially offsets the original increase in drain current.

Figure 13-9b shows the Q point on the transconductance curve. V_{GS} equals

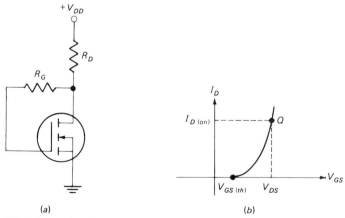

(a)   (b)

Figure 13-9 Drain-feedback bias.

269

Transistor Circuit Approximations

V_{DS}, and the corresponding I_D equals $I_{D(on)}$, a value of drain current well above the threshold point. To assist you, data sheets for E MOSFETs usually include a value of $I_{D(on)}$ for $V_{GS} = V_{DS}$. This helps in setting up the Q point in a drain-feedback–biased circuit.

DC amplifier

A *dc amplifier* is one that can operate all the way down to zero frequency without a loss of gain. One way to build a dc amplifier is the direct-coupled design discussed earlier. Another way is the chopper method.

Figure 13-10 shows a dc amplifier using direct-coupled MOSFETs. The input stage is a D MOSFET with zero bias. The second and third stages use E MOSFETs with gate bias voltages supplied by the drain of the preceding stage. The design of Fig. 13-10 uses MOSFETs with quiescent drain currents of 3 mA. For this reason, each drain runs at 10 V with respect to ground. To get a ground-referenced output, we tap the final output between the 100-kΩ resistors.

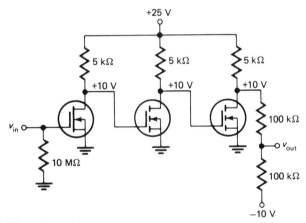

Figure 13-10 DC amplifier using MOSFETs.

The voltage-gain formulas for MOSFETs are identical to those for JFETs; therefore, each stage has a gain of $g_m r_D$. The gain of the output voltage-divider is $\frac{1}{2}$, so that the overall gain is $(g_m r_D)^3/2$.

Other biasing methods

Table 13-1 summarizes biasing methods for JFETs and MOSFETs. Self-bias works with JFETs and D MOSFETs, but not with E MOSFETs, because

270

TABLE 13-1   FET BIASING CIRCUITS

	JFET	D MOSFET	E MOSFET
Self bias	Yes	Yes	No
Current-source bias	Yes	Yes	No
Zero bias	No	Yes	No
Drain-feedback bias	No	No	Yes

V_{GS} is always negative in a self-biased circuit. Also, JFETs and D MOSFETs are fine with current-source bias, but not E MOSFETs. Why? Because a positive V_{GS} would force the bipolar transistor into saturation.

Zero bias is not suitable for JFETs or E MOSFETs, because JFETs need a reverse-biased gate and E MOSFETs need a forward-biased gate. Finally, drain-feedback bias works only with E MOSFETs.

13-4   THE MOSFET AS A DRIVER AND A LOAD

As already mentioned, the E MOSFET has made its greatest impact in digital electronics. MOS integrated circuits are ideal for large-scale integration of computer memories and microprocessors. This section tells you why.

Passive loading

Figure 13-11a shows a MOSFET driver and a *passive load* (an ordinary resistor for R_D). In this switching circuit, v_{in} is either low (less than the threshold voltage) or high (greater than the threshold voltage). Because of this, the MOSFET is either off or on. When v_{in} is low, the MOSFET is cut off and operates at the lower end of the load line (see Fig. 13-11b). In this case, almost all the supply voltage appears across the MOSFET and v_{out} is high.

On the other hand, when v_{in} is high enough, the operating point switches to the upper end of the load line. In this case, v_{out} is low. The drain current through the MOSFET is approximately equal to V_{DD}/R_D.

Active loading

Resistors take up much more chip area than MOSFETs. For this reason, resistors are rarely used in MOS ICs. Figure 13-11c shows an *active load* (another MOSFET taking the place of R_D). This upper MOSFET has its gate returned to the drain; therefore, it's always turned on. Furthermore, this upper MOSFET has a smaller channel and less conductivity than the lower MOSFET. For this reason, the upper MOSFET tends to act like a resistor, whereas the lower MOSFET acts like a switch.

271

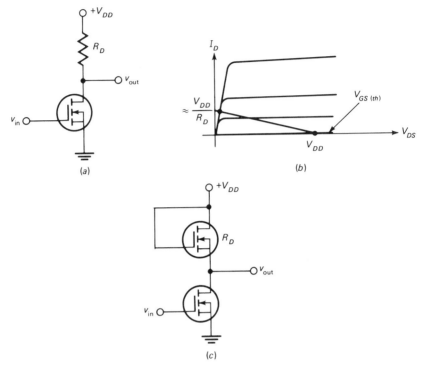

Figure 13-11 *(a)* MOSFET switching. *(b)* Load line. *(c)* Active load.

Using an MOS *driver* (the lower MOSFET) and an MOS load (the upper MOSFET) leads to much smaller integrated circuits than are available with bipolar ICs. This is why MOS technology predominates LSI circuits, especially in the digital field, where manufacturers can fabricate a computer on a chip.

13-5 MOS LOGIC CIRCUITS

p-Channel MOS was one of the first technologies used to build digital ICs. In this approach, *p*-channel E MOSFETs act like switches and active loads. But *p*-channel MOS has a big disadvantage; its carriers are holes instead of free electrons. These holes move more slowly than free electrons, which means the switching speed of a *p*-channel device is not as high as an *n*-channel device. Because of its greater speed, *n*-channel technology has virtually wiped out *p*-channel technology in memory and microprocessor applications.

Another MOS technology called complementary MOS (CMOS) has become important in low-power applications. With CMOS, complementary MOS-

272

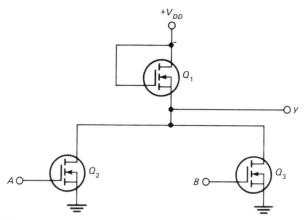

Figure 13-12 NOR gate.

FETs (p-channel and n-channel) are used on the same chip. Because of its lower power consumption, CMOS is the technology used for digital wrist watches and pocket calculators.

This section takes a brief look at some basic MOS logic circuits to give you an idea of how they work.

n-Channel MOS

Figure 13-12 shows a *NOR gate*, one of the basic circuits in digital electronics. The inputs A and B are either low or high voltages; this produces an output y that is either low or high. The upper MOSFET Q_1 acts like a load resistor R_D; the two lower MOSFETs Q_2 and Q_3 act like switches. When A and B are both low, Q_2 and Q_3 are cut off, and the output y is high. This first possibility is shown in Table 13-2.

If A is low and B is high, Q_2 remains cut off but Q_3 turns on, appearing like a closed switch. Therefore, the output drops down to a low value. This is the second possibility of Table 13-2.

TABLE 13-2   NOR GATE

A	B	y
Low	Low	High
Low	High	Low
High	Low	Low
High	High	Low

Transistor Circuit Approximations

The third case is A high and B low. This time, Q_2 closes and pulls the output down to a low value. Finally, when both A and B are high, they both appear as closed switches, resulting in a low output.

A table like Table 13-2 is called a *truth table;* it's nothing more than a summary of all input and output possibilities. As you can see, the truth table for a NOR gate shows that the output is high only when A and B are both low.

Figure 13-13 is another example of n-channel MOS logic. This circuit is called a *NAND gate.* Here's how it works: Q_1 is again an active load resistor;

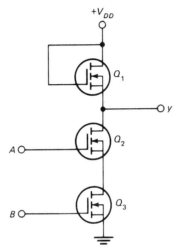

Figure 13-13 NAND gate.

Q_2 and Q_3 are driver switches. When A and B are both low, no current exists and the output y goes to a high value. If A is low and B is high, the open Q_2 driver prevents drain current; therefore, the output y remains high. Similarly, if A is high and B is low, the open Q_3 driver prevents drain current and the output remains high. Finally, when both A and B are high, Q_2 and Q_3 are like closed switches, pulling the output down to a low value.

Table 13-3 summarizes the action of the circuit. As shown in this truth table, a NAND gate has a low output only when both inputs are high.

TABLE 13-3 NAND GATE

A	B	y
Low	Low	High
Low	High	High
High	Low	High
High	High	Low

274

CMOS inverter

Figure 13-14 shows a *CMOS inverter*. Notice that Q_1 is a *p*-channel device and Q_2 is an *n*-channel device. The circuit is analogous to the class B push-pull emitter follower discussed in Chap. 10. In other words, both devices act like switches. When one device is on, the other is off, and vice versa.

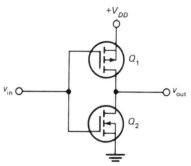

Figure 13-14 CMOS inverter.

For instance, when v_{in} is low, the *n*-channel MOSFET is open, but the *p*-channel MOSFET is closed. For this reason, the output voltage is high. On the other hand, when v_{in} is high, the *n*-channel device is closed, but the *p*-channel device is open. In this case, the output voltage is pulled down to a low value.

The key advantage of CMOS is the low power consumption. Because both devices are in series, the current is determined by the leakage current through the "off" device, which is in nanoamperes. This means the total power dissipation is in nanowatts. This low power dissipation is one of the main reasons for the popularity of CMOS logic circuits.

Table 13-4 summarizes the truth table of the CMOS inverter. As you can see, the output is the *complement* (opposite) of the input. The inverter is sometimes called a *NOT gate* because the output is not the same as the input.

CMOS technology can also be used to build many other logic circuits, including gates (AND, OR, NAND, NOR, etc.) and more advanced circuits.

TABLE 13-4 NOT GATE

v_{in}	v_{out}
Low	High
High	Low

275

13-6 VMOS

Figure 13-15a shows how a conventional E MOSFET appears in an integrated circuit. The source is on the left, the gate in the middle, and the drain on the right. Carriers flow horizontally from the source to the drain when V_{GS}

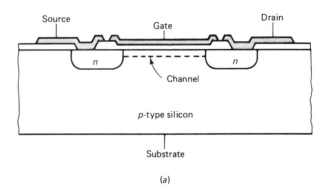

(a)

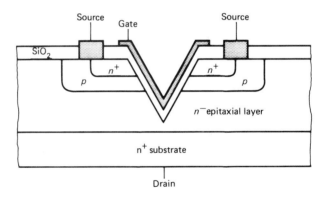

(b)

Figure 13-15 (a) Conventional MOSFET. (b) VMOS.

is greater than the threshold voltage. This conventional structure places a limit on the maximum current, because the carriers must flow in the narrow channel next to the gate. As a result, conventional MOS devices have low power ratings (less than 1 W).

Structure

Figure 13-15b illustrates the structure of *vertical* MOS (VMOS). The two sources are normally connected and the substrate becomes the drain. There-

276

fore, carriers flow vertically. Because of the V-shaped gate, the current path is much wider, which increases the current density. The overall result is an E MOSFET that can handle much larger currents and voltages than a conventional E MOSFET. In other words, VMOS technology has ushered in the high-power MOSFET, one that can compete with bipolar power transistors.

Faster switching speed

One advantage of the VMOS transistor over the bipolar transistor is the lack of charge storage. In a bipolar, the charge storage prevents fast turn-off because stored charges have to be swept out of the base. Until these charges have been removed, the bipolar transistor continues to conduct. Not so with the VMOS transistor. Since it has no charge storage, it can turn off much faster than a bipolar transistor. A VMOS transistor can shut off amperes of current in tens of nanoseconds. This is from 10 to 100 times faster than a comparable bipolar transistor.

No thermal runaway

Another advantage of the VMOS transistor is the lack of *thermal runaway*. With a bipolar power transistor, an increase in junction temperature reduces the base-emitter knee voltage. In turn, this means more collector current flows, causing a further increase in junction temperature. Unless the heatsinking is adequate, the transistor can run away thermally by driving the junction temperature beyond $T_{J(\text{max})}$. This thermal runaway destroys the bipolar transistor. Because the VMOS transistor has no gate current (analogous to base current), it is free of thermal runaway and will not self-destruct like a bipolar.

VMOS transistors are being used in numerous applications requiring high power, such as high-speed switches, audio amplifiers, RF amplifiers, and switching voltage regulators.

GLOSSARY

active load A bipolar transistor or FET that acts like a load resistor for another active device.

dc amplifier One that has no lower cutoff frequency, allowing it to amplify dc as well as ac signals.

inversion layer When the gate-source voltage of an E MOSFET is greater than the threshold voltage, the layer of doped material next to the gate becomes conductive. This conductive layer is the inversion layer.

277

ions Atoms that are either positively or negatively charged.

large-scale integration No standard definition exists, although some people classify LSI as an integrated circuit whose complexity exceeds that of a circuit with 100 gates or logic circuits.

normally off MOSFET Synonymous with an E MOSFET.

normally on MOSFET Synonymous with a D MOSFET.

passive load The same as an ordinary resistor.

thermal runaway A condition where increasing junction temperature reduces the base-emitter knee voltage, which in turn increases the collector current and the junction temperature until $T_{J(max)}$ is exceeded.

threshold voltage The gate-source voltage that just barely produces drain current in an E MOSFET. Below the threshold voltage, drain current is approximately zero.

truth table A table of the input/output possibilities of a logic circuit.

VMOS transistor An E MOSFET with a V-shaped gate. This structure significantly increases the maximum current and power rating of the MOSFET.

SELF-TESTING REVIEW

Read each of the following and provide the missing words. Answers appear at the beginning of the next question.

1. The MOSFET is almost ideal as a switching device because of its low _____ consumption. Furthermore, MOS circuits take up much less chip area than bipolar circuits. This is an important advantage in _____ integration.

2. *(power, large-scale)* The depletion-type MOSFET can operate in either the _____ mode or the _____ mode. The depletion-type MOSFET is called a _____ MOSFET.

3. *(depletion, enhancement, normally on)* The enhancement-type MOSFET is called a _____ MOSFET. The minimum gate-source voltage that creates the *n*-type inversion layer is called the _____ voltage. The newer MOSFETs are protected by built-in _____ diodes.

4. *(normally off, threshold, zener)* Zero bias works only with _____ MOSFETs. Drain-feedback bias is often used with _____ MOSFETs.

5. *(D, E)* A passive load is an ordinary resistor. An _____ load is a

MOSFET taking the place of R_D. Because resistors take up too much chip area, MOSFETs are used in the place of resistors. This leads to much smaller integrated circuits than is possible with bipolar transistors, and explains why MOS technology predominates in _____ integration.

6. *(active, large-scale)* *p*-Channel MOS was one of the first technologies used to build digital ICs. Even better and faster is _____ MOS, which uses free electrons instead of holes as carriers. Because of its low power consumption, _____ is the technology used for digital wrist watches and pocket calculators.

7. *(n-channel, CMOS)* A _____ table is a summary of all the input and output possibilities of a digital circuit. An inverter is sometimes called a _____ gate, because the output is not the same as the input.

8. *(truth, NOT)* The VMOS transistor can handle much larger currents than a conventional MOSFET. This means VMOS transistors are high-power devices that can compete with bipolar power transistors. VMOS transistors have the advantages of faster switching speed and no thermal runaway.

PROBLEMS

13-1. The M113 is a D MOSFET with a gate leakage current of 100 pA when $V_{GS} = 20$ V. Calculate the input resistance.

13-2. A 3N169 is an E MOSFET with a minimum threshold voltage of 0.5 V and a maximum threshold voltage of 1.5 V. If we are going to use thousands of 3N169s, what is the minimum gate voltage that ensures turn-on?

13-3. The 2N3797 has an $I_{DSS(min)}$ of 2 mA, an $I_{DSS(typ)}$ of 2.9 mA, and an $I_{DSS(max)}$ of 6 mA. If this D MOSFET is used in Fig. 13-16*a*, what is the typical value of V_{DS}? The lowest possible V_{DS}?

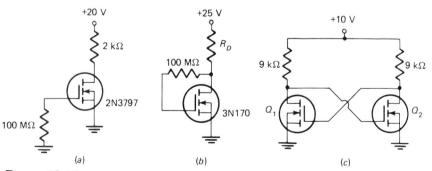

Figure 13-16

Transistor Circuit Approximations

13-4. The 3N170 is an E MOSFET with an $I_{D(\text{on})}$ of 10 mA when $V_{GS} = V_{DS} = 10$ V. Select a value of R_D in Fig. 13-16b that sets up an I_D of 10 mA.

13-5. Figure 13-16c shows part of a digital circuit called a *flip-flop*. The E MOSFETs have threshold voltages of 2 V. When V_{GS} is 10 V, I_D equals 1 mA. Suppose no drain current flows through Q_1.

a. What is the voltage from the drain of Q_1 to ground?
b. How much voltage is applied to the gate of Q_2?
c. How much voltage is there on the drain of Q_2?
d. Why is Q_1 off?

13-6. The D MOSFET of Fig. 13-17a has a g_m of 2000 μS. What is the voltage gain from the gate to the drain?

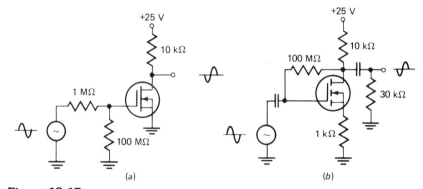

(a)   (b)

Figure 13-17

13-7. The input signal in Fig. 13-17a has a peak of 0.1 V. If the D MOSFET has a g_m of 1500 μS, what is the peak voltage of the ac output signal?

13-8. In Fig. 13-17b, the E MOSFET has a g_m of 2000 μS. What is the voltage gain from the gate to the drain? If the 1-kΩ resistor is replaced by a short circuit, what is the voltage gain?

13-9. In Fig. 13-11a, R_D equals 5 kΩ and V_{DD} is 20 V. What is the output voltage when v_{in} is low? What is the approximate drain current when v_{in} is high?

13-10. The active load of Fig. 13-12 is equivalent to a 100-kΩ resistor. V_{DD} is 10 V. If A and B are low, what is the output voltage? If A is low and B is high, what is the approximate current through the active load?

280

Chapter 14

Operational Amplifiers

About a third of all linear ICs are *operational amplifiers* (op amps). An op amp is a high-gain dc amplifier usable from 0 to over 1 MHz. By connecting external resistors to the op amp, you can adjust the voltage gain and bandwidth to your requirements. There are over 2000 types of commercially available op amps. Almost all are linear ICs with room-temperature dissipations under a watt. Whenever you need voltage gain in a low-power application (less than 1 W), check the available op amps. In many cases, an op amp will do the job.

14-1 GENERATIONS OF OP AMPS

The first stage of an op amp is almost always a differential amplifier and the last stage is usually a class B push-pull emitter follower. Figure 14-1a shows the schematic symbol of an op amp. It has two inputs and one output. The upper input is called the *noninverting* input and is marked with a *plus* sign to indicate that v_{out} is in phase with v_1. The lower input is known as the *inverting* input; it is marked with a *minus* sign to indicate that v_{out} is 180° out of phase with v_2. Since the overall voltage gain of the op amp is A,

$$v_{out} = A(v_1 - v_2) \qquad\qquad (14\text{-}1)$$

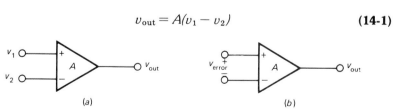

Figure 14-1 *(a)* Op-amp symbol. *(b)* Error voltage.

Transistor Circuit Approximations

This says the output voltage equals the voltage gain times the difference of the two input voltages.

When discussing negative feedback in section 14-5, we will use the *error voltage,* defined as

$$v_{error} = v_1 - v_2$$

This is the voltage between the op amp input terminals as shown in Fig. 14-1*b*. In this case, the output of the op amp is

$$v_{out} = Av_{error} \tag{14-2}$$

The importance of this will be clear later.

First generation

The first quality IC op amp came out in 1965; it was the Fairchild μA709. Similar op amps followed, including Motorola's MC1709, National Semiconductor's LM709, Texas Instruments' SN72709, and others. The last three digits in each of these types are 709. All these op amps behave the same; that is, they all have the same specifications. For this reason, we will refer to them as the 709.

The 709 family has different models like the 709, 709A, 709B, and 709C. For the same manufacturer, all of these have the same schematic diagram, but the tolerances are greater for later model numbers. For example, the 709 has the best tolerance and costs the most. At the other extreme, the 709C has the worst tolerance but costs the least.

The 709C has these typical values: $z_{in} = 250$ kΩ, $z_{out} = 150$ Ω, and $A = 45,000$. These values tell us the 709C has a high input impedance, a low output impedance (the same as Thevenin impedance), and a high voltage gain.

The 709 typifies the first generation of IC op amps. Although widely used in earlier years, the first generation has several disadvantages, including *no short-circuit protection* and the need for *many external components.* No short-circuit protection means the op amp can be accidentally destroyed by shorting the output terminal to ground. Also, having to connect many external components is a nuisance.

Second generation

The 741 is typical of second-generation IC op amps. It has short-circuit protection and fewer external components. Other improvements include its

282

higher input impedance (over a megohm), lower output impedance (75 Ω), and larger voltage gain (200,000). Because it is inexpensive and easy to use, the 741 has become one of the most widely used op amps.

Later generations

The 709 and 741 are industry standards, used for general-purpose applications. For special applications, however, we need newer devices in which one or more specifications have been sharply improved. For instance, if we want high input impedance, *BIFET* op amps are available. These special op amps use JFETs for the input stage, followed by bipolar stages. This combination gives the high input impedance associated with JFETs and the high voltage gain of bipolars. Other improved op amps are available for critical applications requiring low drift, low power consumption, high load current, etc.

Here are the main points to remember about op amps:

1. An op amp uses direct coupling, which gives voltage gain down to zero frequency.
2. It has high input impedance, low output impedance, and high voltage gain.
3. Op amps are rarely used *open-loop;* this means you have to connect a few external components to get normal operation.

14-2 A SIMPLIFIED OP-AMP CIRCUIT

Figure 14-2 is a simplified schematic diagram for a typical op amp. This circuit is equivalent to the 741 and many later-generation op amps. Here are a few basic ideas about how it works.

Input stage

Q_{13} and Q_{14} are a current mirror. Therefore, Q_{14} sources tail current to the input diff amp (Q_1 and Q_2). The diff amp drives a current mirror, consisting of Q_3 and Q_4. An input signal V_{IN} produces an amplified current out of this mirror that goes into the base of Q_5.

Second and third stages

The second stage is an emitter follower *(Q₅)*. It steps up the input impedance of the third stage *(Q₆)* by a factor of β. Note that Q_6 is the driver for the output stage. Incidentally, the plus sign on the Q_5 collector means it's con-

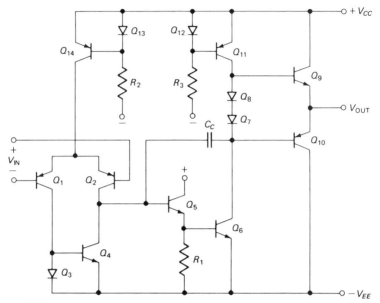

Figure 14-2 Simplified schematic for 741 and similar op amps.

nected to the V_{CC} supply; similarly, the minus signs at the bottom of R_2 and R_3 mean these are connected to the V_{EE} supply.

Output stage

The last stage is a class B push-pull emitter follower (Q_9 and Q_{10}). Because of the split supply, the quiescent output is ideally 0 V. Q_{11} is part of a current mirror that sources current through the compensating diodes (Q_7 and Q_8). Q_{12} is the input half of the mirror, and biasing resistor R_3 sets up the desired mirror current.

Compensating capacitor

C_C is called a *compensating* capacitor. This small capacitor (typically 30 pF) has a pronounced effect on the frequency response. For reasons covered in advanced books, C_C is needed to prevent *oscillations,* unwanted signals produced within the amplifier.

Active loading

In Fig. 14-2, we have another example of active loading. The CE stage (Q_6) drives an active load (Q_{11}). Because Q_{11} is part of a current mirror,

284

it sources a fixed current; therefore, the amplified signal current out of Q_6 is forced into one of the final output transistors (Q_9 or Q_{10}), whichever is conducting.

Input impedance

Recall that the input impedance of a diff amp is

$$z_{in} = 2\beta r'_e$$

With a small tail current in the input diff amp, an op amp can have a fairly high input impedance. For example, the input diff amp of a 741 has a tail current of approximately 15 μA. Since each emitter gets half of this,

$$r'_e = \frac{25 \text{ mV}}{I_E} = \frac{25 \text{ mV}}{7.5 \text{ }\mu\text{A}} = 3.33 \text{ k}\Omega$$

The input transistors have a typical β of 300; therefore,

$$z_{in} = 2\beta r'_e = 2 \times 300 \times 3.33 \text{ k}\Omega = 2 \text{ M}\Omega$$

This agrees with the data-sheet value for a 741.

Summary

The circuit of Fig. 14-2 is relatively simple compared to the actual schematic diagram of an op amp. Nevertheless, this equivalent circuit is a big help in understanding and using op amps. As shown, a split supply drives the typical op amp. Because of the diff-amp input stage, we can apply two input signals. The final output signal is the amplified difference of the two inputs.

14-3 OP-AMP SPECIFICATIONS

The data sheet of an op amp contains all kinds of specifications. What follows is a brief description of some of the more important ones.

Input bias current

The op amp of Fig. 14-2 is equivalent to what is inside the IC package. For the circuit to work properly, you need to connect V_{CC} and V_{EE} supplies, equal in magnitude and opposite in sign. But that's not all. You also have to connect external dc returns for the floating input bases. In other words, the Q_1 and Q_2 base currents have to flow to ground to complete the circuit.

The signal sources driving the op amp provide the dc returns unless they are capacitively coupled; in this case, you have to add resistors from each base to ground. Either way, the base currents of the input diff amp must flow through external resistances to ground. These base currents are close in value, but not necessarily equal. When slightly unequal base currents flow through external dc resistances, they produce a small error voltage or unbalance; this is a false input signal. The smaller the base currents are, the better, because the unbalance is minimized.

The *input bias current* shown on data sheets is the average of the two input currents. It tells you approximately what each input current is. As a guide, the smaller the input bias current, the smaller the possible unbalance. The 741 has an input bias current of 80 nA, which is acceptable in many applications. But in critical applications, a later-generation op amp may be preferred. For example, a BIFET op amp like the 357 has an input bias current of only 30 pA; even if the two input currents are slightly different, the false input signal is small enough to ignore.

Input offset current

The *input offset current* is the difference between the two input currents. The 741 has an input offset current of 20 nA. When working with 741s, we may find 20 nA more current in one base than the other. These unequal base currents produce a false difference voltage that gets amplified to produce a false output voltage. In some applications, this unwanted output voltage is small enough to ignore. The general guide is this: the smaller the input offset current, the better.

Input offset voltage

The *input offset voltage* is the error voltage needed to null or zero the quiescent output voltage of an op amp. For instance, the 741 has a worst-case input offset voltage of 5 mV. When using 741s, we would have to apply a difference input of up to 5 mV to zero the quiescent output voltage.

14-4 SLEW RATE AND POWER BANDWIDTH

Among all specifications affecting the ac operation of an op amp, *slew rate* is the most important because it limits the bandwidth for large signals. Slew rate is defined as the maximum rate of change in output voltage.

286

For instance, the 741 has a slew rate of 0.5 V/μs. This means the output voltage can change no faster than 0.5 V during each microsecond. This is the ultimate speed of a 741.

Step input

Figure 14-3 illustrates the effect of slew rate for a 741. If we overdrive a 741 with a large step input (Fig. 14-3a), the output slews as shown in Fig. 14-3b. It takes 20 μs for the output voltage to change from 0 to 10 V. It is impossible for the 741 output to change faster than this.

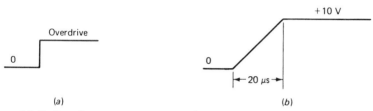

Figure 14-3 Overdriving an op amp produces slew-rate limiting.

Sinusoidal input

Slew-rate limiting can occur even with a sinusoidal signal. Figure 14-4a shows a sinusoidal output with a peak of 10 V. The initial slope of the sine wave where it passes through zero is important. As long as this initial slope is less than the slew rate of the op amp, there is no problem. But when the frequency increases, the initial slope of the sine wave may be greater than the slew rate. In this case, we get the slew-rate distortion shown in Fig. 14-4b. We should get a sinusoidal output (dashed curve); instead, we get a triangular-looking signal because the output voltage cannot change faster than the slew rate. The higher the frequency is, the smaller the swing and the more triangular the output waveform.

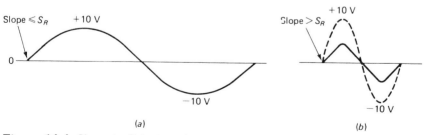

Figure 14-4 Slew-rate distortion of sine wave.

Power bandwidth

Slew-rate distortion of a sine wave starts at the point where the initial slope of the sine wave equals the slew rate of the op amp. With calculus, we can derive this useful formula:

$$f_{max} = \frac{S_R}{2\pi V_P} \qquad\qquad \textbf{(14-3)}$$

where f_{max} = highest undistorted frequency
$\quad S_R$ = slew rate of op amp
$\quad V_P$ = peak of output sine wave

As an example, the slew rate of a 741 is 0.5 V/μs. If the output sine wave is to have a peak of 10 V, the highest undistorted frequency is

$$f_{max} = \frac{S_R}{2\pi V_P} = \frac{0.5 \text{ V}/\mu s}{2\pi \times 10 \text{ V}} = 7.96 \text{ kHz}$$

This means the output will be sinusoidal up to 7.96 kHz; beyond this, the output peak drops off and the shape becomes triangular.

Frequency f_{max} is called the *power bandwidth* of an op amp. We have just found that the 10-V power bandwidth of a 741 is approximately 8 kHz. This means the undistorted bandwidth for large-signal operation is 8 kHz. Try to amplify higher frequencies with the same peak value and you will get slew-rate distortion.

Tradeoff

One way to increase the power bandwidth is to accept less-than-maximum output swing. Figure 14-5 is a graph of Eq. (14-3) for three different slew rates. By trading off output amplitude for frequency, we can improve the power bandwidth. For instance, if an output peak of 1 V is acceptable in an application, the power bandwidth of a 741 increases to approximately 80 kHz (the bottom curve).

The LM318 is an op amp with a slew rate of 50 V/μs. With this high-slew-rate op amp, the 10-V power bandwidth is 800 kHz (top curve). If the output peak can be reduced to 1 V, then the power bandwidth increases to 8 MHz.

14-5 NEGATIVE FEEDBACK

In a feedback control system, the output is sampled and a fraction of it is sent back to the input. This returning signal combines with the original

288

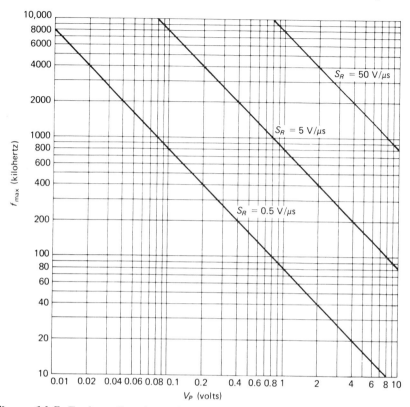

Figure 14-5 Trading off peak amplitude for power bandwidth.

input, producing unusual changes in system performance. *Negative feedback* means the returning signal has a phase that opposes the input signal. The advantages of negative feedback are: stabilizing the gain, improving input and output impedances, reducing nonlinear distortion, and increasing bandwidth. As mentioned earlier, op amps are not meant to be used open-loop; they are intended to be used *closed-loop,* which means connecting external components, usually to get negative feedback.

Basic idea

Figure 14-6 is an example of a negative-feedback amplifier. The input to the internal amplifier (block A) is called the *error voltage*. This voltage is the difference between the input signal v_{in} and the feedback signal Bv_{out}. In a typical negative-feedback system, the error voltage approaches zero; the reason is that the internal gain A is very high, from 10,000 to more than

289

Transistor Circuit Approximations

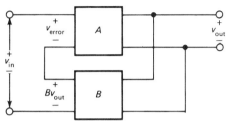

Figure 14-6 Negative feedback.

1,000,000.   Since the output voltage is usually less than 10 V, the error voltage is somewhere in the microvolt region at low frequencies.   Compared to all other voltages in the circuit, v_{error} is insignificant.

Why feedback stabilizes gain

Here's why negative feedback stabilizes the voltage gain.   Suppose the internal gain A increases because of temperature change or some other reason. The output voltage will rise.   This means more voltage is fed back to the input.   Block B is usually a voltage divider whose gain B is between 0 and 1.   The feedback signal Bv_{out} therefore is a fraction of the output voltage. This feedback signal returns with a phase that opposes the input signal v_{in}, causing v_{error} to decrease.   The reduced error voltage to the internal amplifier almost completely offsets the original increase in voltage gain A.   The result is that v_{out} hardly increases at all.

A similar argument applies to a decrease in voltage gain A.   If A decreases for any reason, the output voltage decreases.   In turn, feedback voltage Bv_{out} decreases, causing v_{error} to increase.   This increase in the error voltage almost completely offsets the original decrease in voltage gain A.   As a result, the output voltage shows only the slightest decrease.

This is the basic idea of what negative feedback does: Attempted changes in output voltage are fed back to the input, causing the error voltage to change in the opposite direction.   The overall effect is that output voltage is virtually independent of the changes in the internal gain A, provided A is very large to begin with.

Notice that the output signal is in phase with the input signal.   Therefore, the system is a *noninverting* amplifier with negative feedback.   A later section shows how to use op amps and resistors to build noninverting amplifiers.

Mathematical analysis

In Fig. 14-6, the error voltage is the difference of the input voltage and the feedback voltage:

290

$$v_{\text{error}} = v_{\text{in}} - Bv_{\text{out}} \qquad (14\text{-}4)$$

The output voltage equals the internal gain times the error voltage:

$$v_{\text{out}} = Av_{\text{error}}$$

or with Eq. (14-4), $\qquad v_{\text{out}} = A(v_{\text{in}} - Bv_{\text{out}})$

By rearranging the equation,

$$\frac{v_{\text{out}}}{v_{\text{in}}} = \frac{A}{1 + AB} \qquad (14\text{-}5)$$

This last equation is the voltage gain of the overall system, because it is ratio of v_{out} (the final output) to v_{in} (the input from the source). Here is a crucial point. For the negative feedback to be effective, the product AB must be much greater than 1. When this condition is satisfied, Eq. (14-5) reduces to

$$\frac{v_{\text{out}}}{v_{\text{in}}} \cong \frac{1}{B} \qquad (14\text{-}6)$$

Why is this result so important? Because it says the overall voltage gain no longer depends on the internal gain A, which is temperature- and transistor-dependent. Instead, the overall gain depends only on the value of B. As mentioned earlier, the feedback circuit is usually a voltage divider, a circuit we can make with precision resistors. This means B can be an accurate and stable value. Because of this, the voltage gain of a feedback amplifier becomes a rock-solid value equal to the reciprocal of B.

A final point: The internal gain A is called the *open-loop* gain because it is the gain we would get if the feedback path were opened. On the other hand, the overall gain with feedback is called the *closed-loop* gain because it is the gain we get when there's a closed loop or signal path all the way around the circuit. This is why Eqs. (14-5) and (14-6) often appear as

$$A_{CL} = \frac{A_{OL}}{1 + A_{OL}B} \qquad (14\text{-}7)$$

and $\qquad A_{CL} \cong \dfrac{1}{B} \qquad (14\text{-}8)$

where A_{CL} is the closed-loop gain and A_{OL} is the open-loop gain.

291

Input and output impedance

By a mathematical derivation similar to that just given for voltage gain, we can prove that the input impedance of a noninverting amplifier with negative feedback is given by

$$z_{in(CL)} = (1 + A_{OL}B)z_{in(OL)} \qquad \text{(14-9)}$$

This means that negative feedback *increases* the input impedance. For instance, if the internal amplifier has an input impedance of 1 kΩ and $1 + A_{OL}B$ equals 10,000, then the input impedance of the overall system is

$$z_{in(CL)} = 10,000 \times 1 \text{ k}\Omega = 10 \text{ M}\Omega$$

Another mathematical derivation leads to this expression for the output impedance:

$$z_{out(CL)} = \frac{z_{out(OL)}}{1 + A_{OL}B} \qquad \text{(14-10)}$$

This means negative feedback *reduces* the output impedance by a factor of $1 + A_{OL}B$. In other words, the Thevenin output impedance of the amplifier goes down. If the internal amplifier has a Thevenin impedance of 1 kΩ and $1 + A_{OL}B$ equals 10,000, then

$$z_{out(CL)} = \frac{1 \text{ k}\Omega}{10,000} = 0.1 \text{ }\Omega$$

EXAMPLE 14-1

A noninverting amplifier with negative feedback has $A = 200,000$ and $B = 0.02$. What is the exact closed-loop gain? The approximate closed-loop gain?

SOLUTION

With Eq. (14-7), the exact gain is

$$A_{CL} = \frac{A_{OL}}{1 + A_{OL}B} = \frac{200,000}{1 + (200,000 \times 0.02)} = 49.9875$$

With Eq. (14-8), the approximate gain is

$$A_{CL} \cong \frac{1}{B} = \frac{1}{0.02} = 50$$

Compare the two answers and you can see how close the approximate and exact gains are.

EXAMPLE 14-2

Suppose the internal gain in the preceding example decreases to 100,000. What is the exact value of closed-loop gain?

SOLUTION

$$A_{CL} = \frac{A_{OL}}{1 + A_{OL}B} = \frac{100,000}{1 + (100,000 \times 0.02)} = 49.975$$

The closed-loop gain is still extremely close to 50, despite the open-loop gain dropping in half. This is what negative feedback is all about. You trade off extremely high open-loop gain and its instability for a much lower closed-loop gain that is very stable in value.

This example again illustrates how accurate the approximate formula for closed-loop gain is. In any well-designed negative-feedback amplifier, the closed-loop gain is much smaller than the open-loop gain and closely equals the reciprocal of B. From now on, we will use $A_{CL} \cong 1/B$ for the overall gain of a noninverting amplifier with negative feedback.

14-6 THE OP AMP AS A NONINVERTING AMPLIFIER

Figure 14-7 shows an op amp used as a noninverting amplifier with negative feedback. The input signal drives the noninverting input of the op amp.

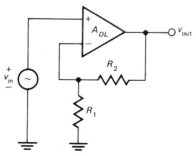

Figure 14-7 Noninverting amplifier.

Transistor Circuit Approximations

The op amp provides internal gain A_{OL}. The external resistors R_1 and R_2 form the feedback voltage divider. Since the returning feedback voltage drives the inverting input, it opposes the input voltage. In other words, the feedback is negative.

Voltage gain

To get a useful formula for voltage gain, we need to find the value of B. The gain of the voltage divider is

$$B = \frac{R_1}{R_1 + R_2}$$

Therefore, the approximate closed-loop gain is

$$A_{CL} \cong \frac{1}{B} = \frac{R_1 + R_2}{R_1}$$

which reduces to
$$A_{CL} \cong \frac{R_2}{R_1} + 1 \qquad\qquad \textbf{(14-11)}$$

This says the closed-loop gain depends on the ratio of the feedback resistors. As already indicated, these can be precision resistors, which means we get a precise value of closed-loop gain. Despite temperature change or op-amp replacement, the closed-loop gain has a rock-solid value.

Upper cutoff frequency

The op amp of Fig. 14-7 has no lower cutoff frequency because it is direct-coupled. But it does have an *open-loop upper cutoff frequency,* designated f_{OL}. The upper cutoff frequency of the overall amplifier is greater than f_{OL} because of the negative feedback. Here's why: When the input frequency increases, we eventually reach the internal cutoff frequency f_{OL}. At this frequency the open-loop gain is down to 0.707 of its maximum gain. Because of the negative feedback, the closed-loop gain shows almost no decrease at all. (Review Example 14-2 if necessary.)

As the input frequency keeps increasing, the open-loop gain keeps decreasing until it approaches the value of closed-loop gain. At this point, the closed-ioop gain starts dropping noticeably. The frequency where the closed-loop gain is down to 0.707 of its maximum value is called the *closed-loop cutoff frequency,* designated f_{CL}. With an advanced derivation, the formula for closed-loop cutoff frequency is

$$f_{CL} = (1 + A_{OL}B)f_{OL} \tag{14-12}$$

where A_{OL} is the maximum open-loop gain. For instance, if f_{OL} is 10 Hz and $1 + A_{OL}B$ is 10,000,

$$f_{CL} = 10,000 \times 10 \text{ Hz} = 100 \text{ kHz}$$

An alternative formula

We can rearrange Eq. (14-7) to get

$$1 + A_{OL}B = \frac{A_{OL}}{A_{CL}}$$

Substituting this into Eq. (14-12) gives

$$f_{CL} = \frac{A_{OL}}{A_{CL}} f_{OL} \tag{14-13}$$

This is very useful because A_{OL}, A_{CL}, and f_{OL} are known for any particular design. For example, a 741C has an A_{OL} of 100,000 and an f_{OL} of 10 Hz. If $A_{CL} = 50$,

$$f_{CL} = \frac{A_{OL}}{A_{CL}} f_{OL} = \frac{100,000}{50} 10 \text{ Hz} = 20 \text{ kHz}$$

This means the overall system has a closed-loop gain of 50 and an upper cutoff frequency of 20 kHz.

Constant gain-bandwidth product

We can rearrange Eq. (14-13) to get

$$A_{CL}f_{CL} = A_{OL}f_{OL} \tag{14-14}$$

The right-hand side of this equation is the product of internal gain and internal cutoff frequency. For instance, the 741C has an A_{OL} of 100,000 and an f_{OL} of 10 Hz. Therefore, the product of its internal gain and cutoff frequency is

Transistor Circuit Approximations

$$A_{OL}f_{OL} = 100{,}000 \times 10 \text{ Hz} = 1 \text{ MHz}$$

This says the internal or open-loop gain-bandwidth product of a 741C equals 1 MHz.

The left-hand side of Eq. (14-14) is the product of closed-loop gain and closed-loop cutoff frequency. No matter what the values of R_1 and R_2 in Fig. 14-7, the product of A_{CL} and f_{CL} must equal the product of A_{OL} and f_{OL}. In other words, the closed-loop gain-bandwidth product equals the open-loop gain-bandwidth product. Given a 741C, the product of A_{CL} and f_{CL} always equals 1 MHz, regardless of the values of R_1 and R_2.

Equation (14-14) is often summarized by saying the *gain-bandwidth product is a constant.* If the right-hand product $A_{OL}f_{OL}$ is constant, so too is the left-hand product $A_{CL}f_{CL}$. For this reason, even though A_{CL} and f_{CL} change when we change external resistors, the product of these two quantities remains constant for a particular op amp.

Unity-gain frequency

Figure 14-8 shows the open-loop response for a typical 741C. The open-loop gain has a maximum value of 100,000. When the operating frequency increases to 10 Hz, the open-loop gain is down to 0.707 of its maximum value. With increasing frequency, the gain keeps dropping off. Well above f_{OL}, the gain decreases by a factor of 10 for each *decade* increase in frequency.

The *unity-gain frequency* is the frequency where the open-loop gain has decreased to unity. In Fig. 14-8, f_{unity} equals 1 MHz. Data sheets often

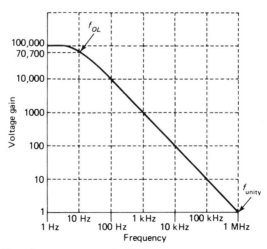

Figure 14-8 Open-loop response of a 741.

296

list the value of f_{unity} because it represents the upper limit on the useful gain of an op amp. The higher f_{unity} is, the better the frequency response of the op amp. For instance, the data sheet of a 741C gives an f_{unity} of 1 MHz, whereas the data sheet of a 318 lists an f_{unity} of 15 MHz. Although it costs more, the 318 gives us usable gain to much higher frequencies than the 741C.

A visual summary

Figure 14-9 summarizes all the key ideas discussed so far. The open-loop gain of a 741C *breaks* at f_{OL}, where it is down to 0.707 of its maximum value. A_{OL} continues decreasing until it approaches the value of closed-loop gain. Then, A_{CL} starts to decrease as previously described. When the frequency equals f_{CL}, the closed-loop gain is down to 0.707 of its maximum value. Thereafter, the curves for A_{OL} and A_{CL} superimpose and decrease to unity at f_{unity}.

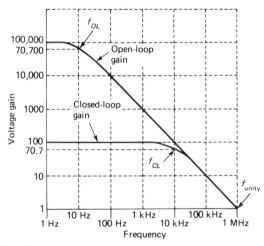

Figure 14-9 Closed-loop and open-loop response.

If the feedback resistors are changed, the closed-loop gain will change to a new value and so too will the closed-loop cutoff frequency. But because the gain-bandwidth product is constant, the closed-loop curve superimposes the open-loop curve beyond the cutoff frequency. For instance, suppose we change the feedback resistors to get a closed-loop gain of 1000. Then, $A_{CL} = 1000$ as shown in Fig. 14-10. For this case, the closed-loop cutoff frequency f_{CL} is 1 kHz.

297

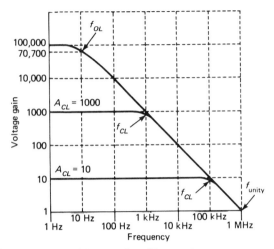

Figure 14-10 Examples of different closed-loop gains.

We can trade off closed-loop gain for bandwidth. For instance, by changing the feedback resistors, we can reduce A_{CL} to 10 as shown in Fig. 14-10. This time, the bandwidth is much greater because f_{CL} is 100 kHz.

Small-signal bandwidth and power bandwidth

Since an op amp has no lower cutoff frequency, the bandwidth equals the upper cutoff frequency. In other words, the open-loop bandwidth is f_{OL}, and the closed-loop bandwidth is f_{CL}. These bandwidths are for small signals where no slew-rate distortion exists. To get the relationship between small-signal bandwidth and power bandwidth, we start by rearranging Eq. (14-3) to get

$$V_P = \frac{S_R}{2\pi f_{max}}$$

Small-signal operation becomes large-signal operation at the point where the small-signal bandwidth f_{CL} equals the power bandwidth f_{max}. Therefore, the maximum output is

$$V_{P(max)} = \frac{S_R}{2\pi f_{CL}} \tag{14-15}$$

As long as the output peak value is less than the $V_{P(max)}$ given by this equation, no slew-rate distortion occurs.

298

EXAMPLE 14-3

Figure 14-11 shows a 741 used as a noninverting amplifier (pin numbers are included). What is the closed-loop voltage gain below the cutoff frequency? What is the small-signal bandwidth? The largest output peak value without slew-rate distortion?

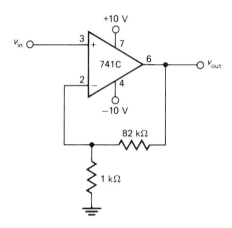

Figure 14-11

SOLUTION

The closed-loop gain is

$$A_{CL} = \frac{R_2}{R_1} + 1 = \frac{82\ \text{k}\Omega}{1\ \text{k}\Omega} + 1 = 83$$

As already given, a 741C has $A_{OL} = 100{,}000$, $f_{OL} = 10$ Hz, and $S_R = 0.5$ V/μs. Therefore, the small-signal bandwidth is

$$f_{CL} = \frac{A_{OL}}{A_{CL}} f_{OL} = \frac{100{,}000}{83}\,(10\ \text{Hz}) = 12\ \text{kHz}$$

With Eq. (14-15), we can calculate the largest output peak without slew-rate distortion:

$$V_{P(\text{max})} = \frac{S_R}{2\pi f_{CL}} = \frac{0.5\ \text{V/μs}}{2\pi(12\ \text{kHz})} = 6.63\ \text{V}$$

As long as you keep the peak output less than 6.63 V, you will have a

299

Transistor Circuit Approximations

closed-loop gain of 83, a closed-loop cutoff frequency of 12 kHz, and no slew-rate distortion.

EXAMPLE 14-4

If R_2 is changed from 82 kΩ to 15 kΩ in Fig. 14-11, what are the new values of A_{CL}, f_{CL}, and $V_{P(max)}$?

SOLUTION

$$A_{CL} = \frac{15 \text{ k}\Omega}{1 \text{ k}\Omega} + 1 = 16$$

$$f_{CL} = \frac{100{,}000}{16}(10 \text{ Hz}) = 62.5 \text{ kHz}$$

$$V_{P(max)} = \frac{0.5 \text{ V/}\mu\text{s}}{2\pi(62.5 \text{ kHz})} = 1.27 \text{ V}$$

Because the small-signal bandwidth has increased from 12 kHz to 62.5 kHz, the maximum allowable peak output decreases from 6.63 V to 1.27 V.

EXAMPLE 14-5

Figure 14-12 shows a *voltage follower*. Calculate its closed-loop voltage gain.

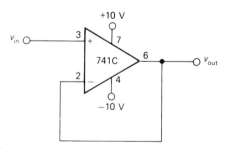

Figure 14-12 Voltage follower.

SOLUTION

R_1 is infinite and R_2 is zero; therefore,

$$A_{CL} = \frac{R_2}{R_1} + 1 = \frac{0}{\infty} + 1 = 1$$

300

This says the closed-loop gain equals unity, which means the output voltage equals the input voltage. Furthermore, the output is in phase with the input. This is why the circuit is called a voltage follower.

Because of the heavy negative feedback, $z_{in(CL)}$ approaches infinity and $z_{out(CL)}$ approaches zero. This makes the voltage follower almost an ideal buffer amplifier since it has very high input impedance, very low output impedance, and a voltage gain of unity. Remember the voltage follower; it's used a lot in industry.

14-7 THE OP AMP AS AN INVERTING AMPLIFIER

Figure 14-13 shows an op amp used as an *inverting* amplifier with negative feedback. The input signal drives the inverting input of the op amp through resistor R_1. The op amp has an open-loop gain of A_{OL}, so that the output signal is much larger than the error voltage. Because of the phase inversion, the output signal is 180° out of phase with the input signal. This means the feedback signal opposes the input signal, and we have negative feedback.

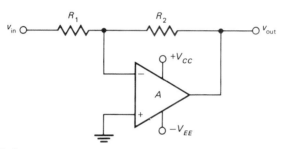

Figure 14-13 Inverting amplifier.

Virtual ground

The open-loop gain of an op amp is extremely large, typically 100,000 for a 741C. If the output voltage is 10 V, the error voltage is only

$$v_{error} = \frac{v_{out}}{A_{OL}} = \frac{10 \text{ V}}{100,000} = 0.1 \text{ mV}$$

Furthermore, the open-loop input impedance of a 741C is around 2 MΩ; for an error voltage of 0.1 mV this means the input current is only

$$i_{error} = \frac{v_{error}}{z_{in(OL)}} = \frac{0.1 \text{ mV}}{2 \text{ MΩ}} = 0.05 \text{ nA}$$

Transistor Circuit Approximations

Because the error voltage is so small compared to all other signal voltages, we can approximate it as *zero*. Likewise, since the error current is so small compared to all other signal currents, we can approximate it also as *zero*.

The inverting input of Fig. 14-13 acts like a *virtual ground*. This means the inverting input appears to be a ground point in the sense that the error voltage is ideally zero for all input voltages. On the other hand, the inverting input is not a true ground point since it has an impedance of 2 MΩ and therefore sinks negligible current. This is why the term *virtual ground* is used; it signifies a point whose voltage with respect to ground is zero, and yet no current can flow into the point.

Voltage gain and input impedance

In Fig. 14-13, the inverting input is a virtual ground. Because of this, all of the input voltage appears across R_1. This sets up a current through R_1 that equals

$$i_1 = \frac{v_{in}}{R_1}$$

All of this current must flow through R_2, because the virtual ground accepts negligible current. Furthermore, the left end of R_2 is ideally grounded, which means all of the output voltage appears across R_2. This allows us to write

$$v_{out} = i_2 R_2 = i_1 R_2 = \frac{v_{in}}{R_1} R_2$$

which rearranges into

$$A_{CL} = \frac{R_2}{R_1} \qquad\qquad \textbf{(14-16)}$$

Also, the input impedance seen by the source is

$$z_{in(CL)} = R_1 \qquad\qquad \textbf{(14-17)}$$

One reason for the popularity of the inverting amplifier is this: It allows us to set up a precise value of input impedance as well as voltage gain. There are many applications where we want to pin down the input impedance, along with the voltage gain. Despite temperature change or op-amp replacement, an inverting amplifier like Fig. 14-13 has a constant input impedance and voltage gain.

302

More than one input

Another advantage of the inverting amplifier is its ability to handle more than one input at a time, as shown in Fig. 14-14a. Because of the virtual ground,

$$i_1 = \frac{v_1}{R_1}$$

$$i_2 = \frac{v_2}{R_2}$$

and

$$v_{\text{out}} = (i_1 + i_2)R_3 = i_1 R_3 + i_2 R_3$$

or

$$v_{\text{out}} = \frac{R_3}{R_1} v_1 + \frac{R_3}{R_2} v_2 \qquad \textbf{(14-18)}$$

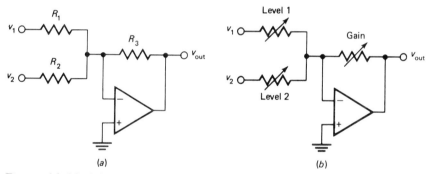

(a) (b)

Figure 14-14 *(a)* More than one input. *(b)* Mixing inputs.

This means we can have a different gain for each input; the output is the sum of the amplified inputs. The same idea applies to any number of inputs; add another input resistor for each new input signal.

Figure 14-14b shows a convenient way to additively mix two signals. The adjustable input resistors allow us to set the level of each input, and the gain control lets us control the output. A circuit like this is useful for combining audio signals.

Small-signal bandwidth and power bandwidth

Like the noninverting amplifier discussed earlier, the inverting amplifier has a constant gain-bandwidth product. The small-signal bandwidth is

303

$$f_{CL} = \frac{A_{OL}}{A_{CL}} f_{OL} \qquad \text{(14-19)}$$

Also, to avoid slew-rate distortion, the maximum allowable output peak is given by

$$V_{P(\max)} = \frac{S_R}{2\pi f_{CL}} \qquad \text{(14-20)}$$

14-8 OP-AMP APPLICATIONS

IC op amps are inexpensive, versatile, and easy to use. For this reason, they are used not only for negative-feedback amplifiers, but also for waveshaping, filtering, and mathematical operations. This section discusses some common applications of op amps.

Comparators

The simplest way to use an op amp is open-loop (no feedback resistors), as shown in Fig. 14-15a. Because of the high gain of the op amp, the slightest error voltage (typically in microvolts) produces maximum output swing. When V_1 is greater than V_2, the error voltage is positive and the output voltage goes to its maximum positive value, typically 1 to 2 V less than the supply voltage. On the other hand, if V_1 is less than V_2, the output voltage swings to its maximum negative value.

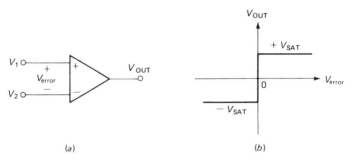

Figure 14-15 Comparator. *(a)* Circuit. *(b)* Response.

Figure 14-15b summarizes the action. A positive error voltage drives the output to $+V_{\text{SAT}}$, the maximum positive value of output voltage. A negative error voltage produces an output of $-V_{\text{SAT}}$. When an op amp is used like

304

this, it's called a *comparator* because all it does is compare V_1 to V_2, producing a saturated positive or negative output, depending on whether V_1 is greater or less than V_2.

Figure 14-16*a* is one application of a comparator: a *go-no go* detector. When V_1 exceeds V_2, the output goes positive and turns on the green LED; this indicates the *go* condition. On the other hand, if V_1 is less than V_2, the output swings negative and turns on the red LED, indicating *no-go*. By selecting proper values for R_1 and R_2, we can set up any V_2. In this way, we can detect when V_1 is greater or less than a desired voltage level.

Figure 14-16*b* shows another application: producing square waves from

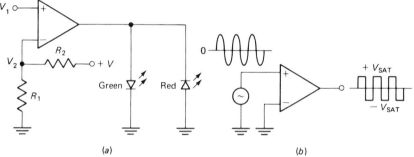

(a)   (b)

Figure 14-16 *(a)* Go-no detector. *(b)* Squaring circuit.

sine waves. Since the inverting input is grounded, V_2 is zero. Therefore, when the input sine wave goes slightly positive, the output saturates positively. When the input goes slightly negative, the output saturates negatively. This is why the input sine wave produces a square-wave output.

Current sources

At times we want to source a fixed current through a load. Figure 14-17*a* shows one way to do it. Since the error voltage is negligibly small, essentially all of V_{IN} appears across R, producing a current of

$$I_{OUT} = \frac{V_{IN}}{R}$$

All this current must flow through the load, because negligible current flows into the op-amp inverting input. Depending on the application, the load may be a resistor, capacitor, inductor, or combination.

Transistor Circuit Approximations

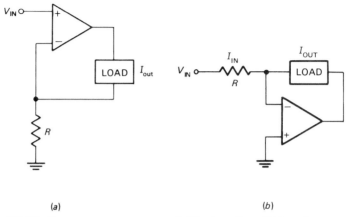

(a)   (b)

Figure 14-17 Op-amp current sources. *(a)* Noninverting. *(b)* Inverting.

Figure 14-17*b* shows an inverting amplifier used to source current through a load. Because of the virtual ground,

$$I_{\text{OUT}} = \frac{V_{\text{IN}}}{R}$$

This allows us to set up a precise value of I_{OUT}. Again, the load may be a resistor, capacitor, inductor, or some combination.

Active half-wave rectifier

Op amps can enhance the performance of the diode circuits discussed in Chap. 4. For one thing, an op amp eliminates the effect of diode offset voltage, allowing us to rectify, peak-detect, clip, and clamp low-level signals (those with amplitudes less than 0.7 V).

Figure 14-18*a* is an *active half-wave rectifier,* one that includes an op amp. When the input signal goes positive, the output goes positive and turns on the diode. The circuit then acts like a voltage follower, and the positive half cycle appears across the load resistor. On the other hand, when the input goes negative, the op-amp output goes negative and turns off the diode. Since the diode is open, no voltage appears across the load resistor. This is why the output is almost a perfect half-wave signal.

The high gain of the op amp virtually eliminates the effect of diode offset voltage. For instance, if the offset voltage ϕ is 0.7 V and the open-loop gain A is 100,000, the input that just turns on the diode is

306

$$V_{IN} = \frac{0.7 \text{ V}}{100,000} = 7 \ \mu V$$

When the input is greater than 7 μV, the diode turns on. The effect is equivalent to reducing the offset potential by a factor of A. In symbols,

$$\phi' = \frac{\phi}{A} \tag{14-21}$$

where ϕ' is the offset potential seen by the input signal. This means we can rectify low-level signals.

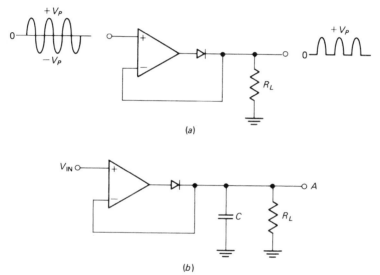

(a)

(b)

Figure 14-18 *(a)* Active half-wave rectifier. *(b)* Active peak detector.

Other active diode circuits

To peak-detect small signals, we can use an *active peak detector* like Fig. 14-18*b*. Because of the op-amp gain, the input offset potential ϕ' is in the microvolt region. This means the circuit can peak-detect signals whose amplitudes are much less than a volt. Furthermore, when the diode is on, the heavy negative feedback produces a Thevenin output impedance approaching zero. For this reason, the charging time constant shrinks to a very small value, eliminating source effects.

Figure 14-19*a* illustrates an *active positive clipper*. With the wiper all

Transistor Circuit Approximations

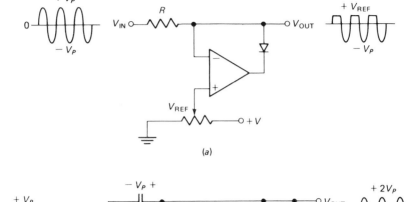

(a)

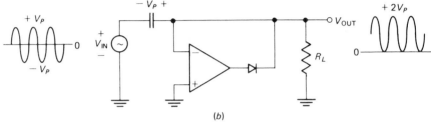

(b)

Figure 14-19 *(a)* Active clipper. *(b)* Active clamper.

the way to the left, V_{REF} is zero and the noninverting input is grounded. When V_{IN} goes positive, the error voltage drives the op-amp output negative and turns on the diode. This means the final output V_{OUT} is at virtual ground for any positive value of V_{IN}. When V_{IN} goes negative, the op-amp output is positive, which turns off the diode and opens the loop. As this happens, the virtual ground is lost, and the final output V_{OUT} is free to follow the negative half cycle of input voltage. This is why the negative half cycle appears at the output. By adjusting V_{REF}, we can change the clipping level.

Figure 14-19b is an *active positive clamper*. The first negative half cycle produces a positive op-amp output that turns on the diode. This allows the capacitor to charge to the peak value of the input. Just beyond the negative peak, the diode turns off, the loop opens, and the virtual ground is lost. Since V_P is being added to a sinusoidal input voltage, the final output waveform is shifted positively through V_P volts. In other words, we get the positive clamped output shown. It swings from 0 to $2V_P$. Because the input offset potential is in the microvolt region, the circuit can positively clamp low-level signals.

Active filters

Virtual ground, an unusually powerful feature, is the reason for the inverting amplifier's versatility. Figure 14-20a shows how virtual ground is used to

308

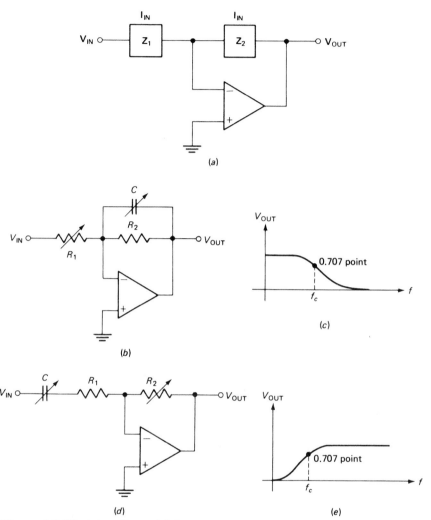

Figure 14-20 Active filters. *(a)* General case. *(b)* Low-pass circuit. *(c)* Low-pass response. *(d)* High-pass circuit. *(e)* High-pass response.

build *active filters.* In this circuit, the input and output voltages are phasors; $\mathbf{Z}_1$ and $\mathbf{Z}_2$ are complex impedances. The virtual ground allows us to write

$$\mathbf{V}_{IN} = \mathbf{I}_{IN}\mathbf{Z}_1$$

and

$$\mathbf{V}_{OUT} = -\mathbf{I}_{IN}\mathbf{Z}_2$$

Transistor Circuit Approximations

where the minus sign indicates phase inversion. Taking the ratio of output voltage to input voltage gives the voltage gain:

$$\frac{V_{OUT}}{V_{IN}} = -\frac{Z_2}{Z_1} \qquad \textbf{(14-22)}$$

This says the voltage gain equals the ratio of two external impedances. By properly selecting these impedances, we can build active filters.

Figure 14-20b is an *active low-pass filter.* At low frequencies the capacitor appears open, and the circuit acts like an inverting amplifier with a voltage gain of $-R_2/R_1$ (the minus sign indicates phase inversion). As the frequency increases, the capacitive reactance decreases, causing the voltage gain to drop off.

Figure 14-20c illustrates the output response. The output signal is maximum at low frequencies. When the frequency reaches the cutoff frequency, the output is down to 0.707 times its low-frequency value. By applying Eq. (14-22), we can derive this formula for the cutoff frequency:

$$f_c = \frac{1}{2\pi R_2 C} \qquad \textbf{(14-23)}$$

The adjustable C of Fig. 14-20b controls the cutoff frequency, and the adjustable R_1 varies the voltage gain.

The *active high-pass filter* of Fig. 14-20d is different. At low frequencies the capacitor appears open, and the voltage gain approaches zero. At high frequencies the capacitor appears shorted, and the circuit becomes an inverting amplifier with a voltage gain of $-R_2/R_1$. Figure 14-20e is the response. By taking the ratio of impedances, we can derive this formula for the cutoff frequency:

$$f_c = \frac{1}{2\pi R_1 C} \qquad \textbf{(14-24)}$$

GLOSSARY

BIFET op amp This is an IC op amp using JFETs for the input stage and bipolar transistors for the remaining stages.

closed-loop gain The voltage gain of a feedback amplifier when there's a closed loop or signal path all the way around the circuit.

error voltage This is the voltage between the noninverting and inverting

310

inputs of an op amp.   Because of the high gain, the error voltage approaches zero.

gain-bandwidth product   This is a figure of merit for op amps, a way to compare their frequency response.   It equals the unity-gain frequency.

negative feedback   Whenever the feedback signal has a phase that opposes the input signal, you have negative feedback.

op amp   A high-gain direct-coupled amplifier usable from 0 to over 1 MHz. It has high input impedance and low output impedance.   It is not meant to be used open-loop.

open-loop gain   The voltage gain of a feedback amplifier when the feedback path is opened.   It is the same as the voltage gain of the internal amplifier.

power bandwidth   For a specified output voltage, it is the highest frequency an op amp can handle without slew-rate distortion.

slew rate   This is the maximum rate of output voltage change for an op amp.

unity-gain frequency   The frequency where the open-loop gain equals unity.   This frequency equals the gain-bandwidth product of the op amp.

virtual ground   Any point in a circuit where the voltage with respect to ground ideally approaches zero, and yet no current can flow into the point. The inverting input of an op amp is a virtual ground when the noninverting input is grounded.

voltage follower   A noninverting op amp with negative feedback and a closed-loop gain of unity.

SELF-TESTING REVIEW

Read each of the following and provide the missing words.   Answers appear at the beginning of the next question.

1. An op amp has a noninverting input and an _____ input.   The _____ voltage appears between these two inputs.

2. *(inverting, error)* The 709 is an example of the _____ generation of IC op amps.   The _____ is typical of the second generation and has become one of the most widely used op amps.   BIFET op amps are later-generation devices that use _____ for the input stage and bipolar transistors for later stages.

311

Transistor Circuit Approximations

3. *(first, 741, JFETs)* The lower that input bias current and input offset current are, the less the unbalance. The input _____ voltage is the error voltage needed to zero or null the output.

4. *(offset)* Slew rate is the maximum _____ of output voltage change. We can get slew-rate disortion even with sinusoidal signals. The power _____ equals the highest frequency the op amp can handle without slew-rate distortion. Power bandwidth increases when output voltage _____.

5. *(rate, bandwidth, decreases)* The internal gain is called the open-loop gain, whereas the gain of the overall feedback amplifier is the _____ gain. With a noninverting amplifier, negative feedback decreases the gain, _____ the input impedance, and _____ the output impedance.

6. *(closed-loop, increases, decreases)* The gain-bandwidth product is constant for a given op amp. For this reason, the product of A_{CL} and _____ is constant even though we change the feedback resistors. The unity-gain frequency is the frequency where the open-loop gain equals _____.

7. *(f_{CL}, unity)* A voltage follower has a closed-loop gain of _____ and the output signal is in _____ with the input signal. This makes the voltage follower almost an ideal buffer.

8. *(unity, phase)* The inverting amplifier with negative feedback has a virtual ground point at the inverting input. This type of ground point has ideally zero voltage to ground. Ideally, no current can flow into this point.

PROBLEMS

14-1. A 741C has a typical voltage gain of 100,000. What is the error voltage if the output voltage is 2 V?

14-2. An op amp has a slew rate of 1 V/μs. What is the power bandwidth for a peak output voltage of 3 V?

14-3. A 318 has a slew rate of 50 V/μs. What is the power bandwidth when the output peak voltage is 8 V?

14-4. A noninverting amplifier with negative feedback has an open-loop gain A_{OL} of 250,000 and a feedback factor B of 0.005. The open-loop input impedance is 2 kΩ and the open-loop output impedance is 500 Ω. Calculate the values of A_{CL}, $z_{in(CL)}$, and $z_{out(CL)}$.

312

14-5. The noninverting amplifier of Fig. 14-7 has $R_2 = 330$ kΩ and $R_1 = 1.5$ kΩ. What does the closed-loop gain equal?

14-6. An op amp has $A_{OL} = 200,000$ and $f_{OL} = 8$ Hz. Calculate the closed-loop cutoff frequency if $A_{CL} = 400$.

14-7. The op amp used in a noninverting amplifier has a gain-bandwidth product of 2 MHz. What does f_{CL} equal when A_{CL} is 700?

14-8. An op amp has a slew rate of 5 V/μs. What is the maximum output peak voltage without slew-rate distortion if the closed-loop cutoff frequency is 200 kHz?

14-9. In Fig. 14-13, $R_1 = 2$ kΩ and $R_2 = 68$ kΩ. What is the closed-loop input impedance? The closed-loop voltage gain?

Chapter 15

Thyristors

A *thyristor* is a special kind of semiconductor switch that uses internal feedback to produce *latching* action. Unlike bipolars and FETs, which can operate either as linear amplifiers or switches, thyristors can only operate in the switching mode. Their main application is in controlling large amounts of load power in motors, heaters, lighting systems, etc.

15-1 THE IDEAL LATCH

All thyristor devices can be explained in terms of the ideal *latch* shown in Fig. 15-1a. Notice that the upper transistor Q_1 is a *pnp* device, and the lower transistor Q_2 is *npn*.

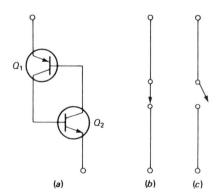

(a)   (b)   (c)

Figure 15-1 Complementary latch.

314

Regeneration

Because of the unusual connection in Fig. 15-1*a*, we have *positive feedback,* also called *regeneration.* A change in current at any point in the loop is amplified and returned to the starting point with the same phase. For instance, if the Q_2 base current increases, the Q_2 collector current increases. This forces more base current through Q_1. In turn, this produces a larger Q_1 collector current, which drives the Q_2 base harder. This build-up in currents will continue until both transistors are driven into saturation. In this case, the latch acts like a closed switch (Fig. 15-1*b*).

On the other hand, if something causes the Q_2 base current to decrease, the Q_2 collector current will decrease. This reduces the Q_1 base current. In turn, there is less Q_1 collector current, which reduces the Q_2 base current even more. This regeneration continues until both transistors are driven into cutoff. At this time, the latch acts like an open switch (Fig. 15-1*c*).

The latch can be in either of two states, closed or open. It will remain in either state *indefinitely.* If closed, it stays closed until something causes the currents to decrease. If open, it stays open until something else forces the currents to increase.

Triggering

One way to close a latch is by *triggering,* applying a forward-bias voltage to either base. For example, Fig. 15-2*a* shows a *trigger* (sharp pulse) hitting the Q_2 base. Suppose the latch is open before point *A* in time. Then, all

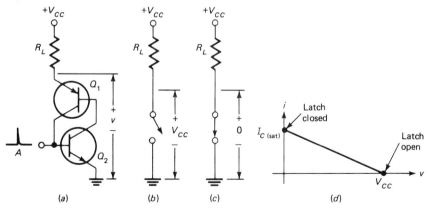

Figure 15-2 Latch action. *(a)* Circuit. *(b)* Open latch. *(c)* Closed latch. *(d)* Load line.

315

the supply voltage appears across the open latch (Fig. 15-2b), and the operating point is at the lower end of the load line (Fig. 15-2d).

At point A in time, the trigger momentarily forward-biases the Q_2 base. The Q_2 collector current suddenly comes on and forces base current through Q_1. In turn, the Q_1 collector current comes on and drives the Q_2 base harder. Since the Q_1 collector now supplies the Q_2 base current, the trigger pulse is no longer needed. In other words, once the regeneration starts, it will sustain itself and drive both transistors into saturation. The minimum input current needed to start the regenerative switching action is called the *trigger current.*

When saturated, both transistors ideally look like short circuits and the latch is closed (Fig. 15-2c). Ideally, the latch has zero voltage across it when closed, and the operating point is at the upper end of the load line (Fig. 15-2d).

Breakover

Another way to close a latch is by *breakover.* This means using a large enough supply voltage V_{CC} to break down either collector diode. Once the breakdown begins, current comes out of one of the collectors and drives the other base. The effect is the same as if the base had received a trigger. Although breakover starts with a breakdown of one of the collector diodes, it ends with both transistors in the saturated state. This is why the term *breakover* is used instead of breakdown to describe this kind of latch closure.

Low-current dropout

How do we open an ideal latch? One way is to reduce the load current to zero; this forces the transistors to come out of saturation and return to the open state. For instance, in Fig. 15-2a we can open the load resistor. Alternatively, we can reduce the V_{CC} supply to zero. In either case, a closed latch will be forced to open. We call this type of opening *low-current dropout* because it depends on reducing latch current to a low value.

Reverse-bias trigger

Another way to open the latch would be to apply a reverse-bias trigger in Fig. 15-2a. For instance, if a negative trigger is used instead of a positive one, the regeneration will drive the transistors into cutoff, and the latch is then open. As an alternative, we can drive the base of Q_1 with a positive trigger. This will reduce the Q_1 base current and start the regeneration that eventually opens the latch.

316

Summary

Here is a summary that will help you understand how different thyristors work:

1. We can close an ideal latch by forward-bias triggering or by breakover.
2. We can open an ideal latch by reverse-bias triggering or by low-current dropout.

Some of the simpler thyristors rely on breakover to close the latch and low-current dropout to open the latch. More complex thyristors use forward-bias triggering to close the latch and reverse-bias triggering to open the latch.

15-2 THE FOUR-LAYER DIODE

Figure 15-3a is a *four-layer diode* (also called a Shockley diode). It is classified as a diode because it has only two external leads. Because of its four doped regions, it's often called a *pnpn* diode. The easiest way to understand how it works is to visualize it separated into two halves as shown in Fig. 15-3b. The left half is a *pnp* transistor and the right half is an *npn* transistor. Therefore, the four-layer diode is equivalent to the latch shown in Fig. 15-3c.

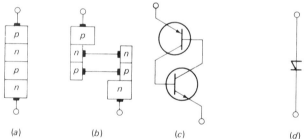

(a)   (b)   (c)   (d)

Figure 15-3 Four-layer diode. *(a)* Structure. *(b)* Equivalent structure. *(c)* Equivalent circuit. *(d)* Schematic symbol.

Because there are no trigger inputs, the only way to close a four-layer diode is by breakover, and the only way to open it is by low-current dropout. With a four-layer diode it is not necessary to reduce the current all the way to zero to open the latch. The internal transistors of the four-layer diode will come out of saturation when the current is reduced to a low value called the *holding current.* Figure 15-3d shows the schematic symbol of a four-layer diode.

317

Transistor Circuit Approximations

EXAMPLE 15-1

The 1N5158 of Fig. 15-4a has a breakover voltage of 10 V. What is the load current when the input voltage equals 15 V and the diode drop is 1 V?

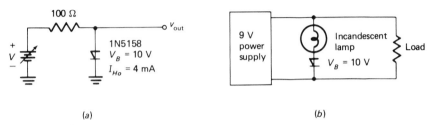

(a)   (b)

Figure 15-4

SOLUTION

There is more than enough voltage to cause breakover. The load current is

$$I = \frac{15\ V - 1\ V}{100\ \Omega} = 140\ mA$$

EXAMPLE 15-2

In Fig. 15-4a, the diode has a holding current of 4 mA. Allowing 0.5 V across the diode at the dropout point, what is the input voltage that just produces low-current dropout?

SOLUTION

To open the four-layer diode, we have to reduce the current below the holding current of 4 mA. This means reducing the input voltage to slightly less than

$$V = 0.5\ V + (4\ mA \times 100\ \Omega) = 0.9\ V$$

EXAMPLE 15-3

Describe the action in Fig. 15-4b.

SOLUTION

The four-layer diode has a breakover voltage of 10 V. As long as the power supply puts out 9 V, the four-layer diode is open and the lamp is dark. But if something goes wrong with the power supply and its voltage rises above 10 V, the four-layer diode latches and the lamp comes on. Even if

318

the supply should return to 9 V, the diode remains latched as a record of the overvoltage that occurred. The only way to make the lamp go out is by turning off the supply.

The circuit is an example of an *overvoltage detector.* As long as the supply stays within normal limits, nothing happens. But if we get an overvoltage, even temporarily, the lamp comes on and stays on.

EXAMPLE 15-4

Figure 15-5a shows a *sawtooth generator.* Describe the circuit action.

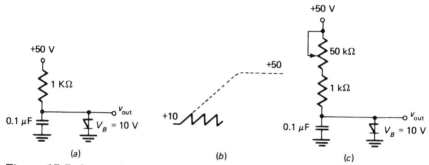

Figure 15-5 Sawtooth generator. *(a)* Fixed frequency. *(b)* Variable frequency.

SOLUTION

If the four-layer diode were not in the circuit, the capacitor would charge exponentially and its voltage would follow the dashed curve of Fig. 15-5b. But the four-layer diode is in the circuit; therefore, as soon as the capacitor voltage reaches 10 V, the diode breaks over and the latch closes. This discharges the capacitor, producing the *flyback* (sudden decrease) of capacitor voltage. At some point on the flyback, the current drops below the holding current and the four-layer diode opens. The next cycle then begins.

Figure 15-5a is an example of a *relaxation oscillator,* a circuit that generates an output signal whose frequency depends on the charging and discharging of a capacitor (or inductor). If we increase the RC time constant, the capacitor takes longer to charge to 10 V and the frequency of the sawtooth wave is lower. For instance, using a rheostat as shown in Fig. 15-5c, we can get a 50:1 range in frequency.

15-3 THE SILICON-CONTROLLED RECTIFIER

The *silicon-controlled rectifier* (SCR) is far more useful than a four-layer diode because it has an extra lead connected to the base of the *npn* section,

319

as shown in Fig. 15-6*a.* We can again visualize the four doped regions separated into two transistors as shown in Fig. 15-6*b.* Therefore, the SCR is equivalent to a latch with a trigger input (Fig. 15-5*c*). Schematic diagrams use the symbol of Fig. 15-5 *d;* whenever you see this, remember it's a latch with a trigger input.

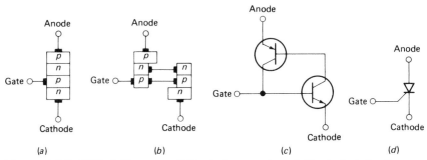

Figure 15-6 SCR. *(a)* Structure. *(b)* Equivalent structure. *(c)* Equivalent circuit. *(d)* Schematic symbol.

Blocking voltage

SCRs are not intended for breakover operation. Breakover voltages are from around 50 V to more than 2500 V, depending on the SCR type number. Most SCRs are designed for trigger closing and low-current opening. In other words, the SCR stays open until a trigger hits the *gate* (see Fig. 15-6 *d*). Then, the SCR latches and remains closed, even though the trigger disappears. The only way to open an SCR is with low-current dropout.

Most people think of the SCR as a device that blocks voltage until a trigger closes it. For this reason, the breakover voltage is often called the *forward blocking voltage* on data sheets. For instance, a 2N4444 is an SCR with a forward blocking voltage of 600 V. As long as the supply voltage is less than 600 V, the SCR cannot break over; the only way to close it is with a gate trigger.

High currents

Almost all SCRs are industrial devices that can handle large currents ranging from less than 1 A to more than 2500 A, depending on the type. Because they are high-current devices, SCRs have relatively large trigger and holding currents. For instance, the 2N4444 can conduct up to 8 A continuously; its trigger current is 10 mA, and so too is its holding current. This means you have to supply the gate with at least 10 mA to control up to 8 A of

320

anode current. (The anode and cathode are shown in Fig. 15-6 d.) As another example, the C701 is an SCR that can conduct up to 1250 A with a trigger current of 150 mA and a holding current of 500 mA.

Critical rate of rise

In many applications, an ac supply voltage is used with the SCR. By triggering the gate at a certain point in the cycle, we can control large amounts of ac power to a load such as a motor, a heater, etc. Because of junction capacitances inside the SCR, it is possible for a rapidly changing supply voltage to trigger the SCR. In other words, if the rate of rise of forward voltage is high enough, capacitive charging currents can initiate regeneration and reduce the breakover voltage.

To avoid false triggering of the SCR, the anode rate of voltage change must not exceed the *critical rate of voltage rise* listed on the data sheet. For instance, a 2N4444 has a critical rate of voltage rise of 50 V/μs. To avoid false breakover, the anode voltage must not rise faster than 50 V/μs. As another example, the C701 has a critical rate of voltage rise of 200 V/μs. To avoid a false closure, the anode voltage must not increase faster than 200 V/μs.

Switching *transients* are the main cause of exceeding the critical rate of voltage rise. One way to reduce the effects of transients is with an *RC snubber,* shown in Fig. 15-7 a. If a high-speed transient does appear on the supply, its rate of rise is reduced at the anode because of the *RC* network. The rate of rise depends on the load resistance, as well as the R and C values.

Larger SCRs also have a *critical rate of current rise.* For instance, the C701 has a critical rate of 150 A/μs. If the anode current tries to rise

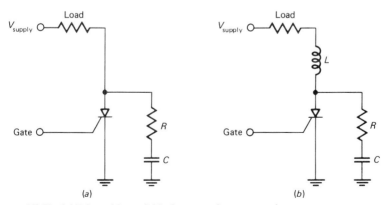

Figure 15-7 *(a)* RC snubber. *(b)* Inductor reduces rate of current rise.

321

Transistor Circuit Approximations

faster than this, hot spots can occur inside the SCR and destroy it. Including an inductor in series as shown in Fig. 15-7 *b* reduces the rate of current rise, as well as helping the *RC* snubber decrease the rate of voltage rise.

EXAMPLE 15-5

Figure 15-8 shows an *SCR crowbar*, a way of protecting a load against overvoltage. The 2N4441 has a forward blocking voltage of 50 V. Describe the circuit action.

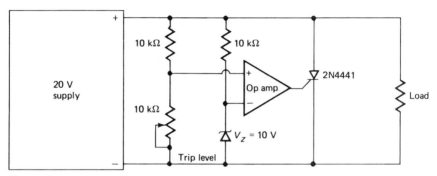

Figure 15-8

SOLUTION

The SCR is normally open because the supply voltage is only 20 V; therefore, the voltage across the load is 20 V. Because of the zener diode, 10 V goes to the inverting input of the op amp. The *trip level* is adjusted to get slightly less than 10 V going into the noninverting input. As a result, the error voltage is negative and the op amp has a negative output. This negative output does nothing to the SCR.

If something goes wrong with the power supply and its voltage tries to rise above 20 V, the noninverting input becomes greater than 10 V. In this case, the error voltage is positive and the op amp delivers a positive trigger to the gate. Immediately, the SCR closes and shuts down the power supply. The action is the same as throwing a crowbar across the load terminals. Because the SCR turn-on is very fast (1 μs for the 2N4441), the load is quickly protected from the damaging effects of a large overvoltage.

Crowbarring is a drastic form of protection, but it is necessary with many digital ICs; they can't take much overvoltage. Rather than risk overvoltage destruction of expensive ICs, therefore, we can use an SCR crowbar to shut down the power supply at the first sign of overvoltage.

15-4 VARIATIONS OF THE SCR

Other *pnpn* devices exist whose action is similar to the SCR. What follows is a brief description of these SCR variations. The devices to be discussed are for low-power applications.

LASCR

The *light-activated* SCR (LASCR) is shown in Fig. 15-9*a*. The arrows represent incoming light that passes through a window and hits the depletion layers. When the light is strong enough, bound electrons can be dislodged from valence orbits to become free electrons. When these free electrons flow out of a collector and into a base, regeneration starts and the LASCR closes.

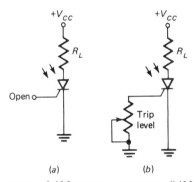

Figure 15-9 LASCR circuits. *(a)* Maximum sensitivity. *(b)* Variable trip point.

After a *light trigger* has closed the LASCR, it remains closed even though the light disappears. For maximum sensitivity to light, the gate is left open as shown in Fig. 15-9*a*. If you want an adjustable trip level, you can include an adjustment as shown in Fig. 15-9*b*. This gate resistor diverts some of the light-produced electrons and changes the sensitivity of the circuit to the incoming light.

GCS

As mentioned earlier, low-current dropout is the normal way to open an SCR. But the *gate-controlled switch* (GCS) is designed for easy opening with a reverse-biased trigger. In other words, a GCS is closed by a positive trigger and opened by a negative trigger (or by low-current dropout). Figure

323

Transistor Circuit Approximations

15-10 shows a GCS circuit. Each positive trigger closes it, and each negative trigger opens it. Because of this, we get the square-wave output shown. The GCS is useful in counting circuits, digital circuits, and other applications where a negative trigger is available for turn-off.

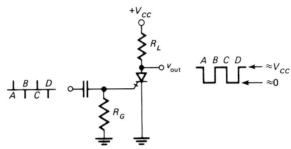

Figure 15-10 GCS circuit.

SCS

Figure 15-11a shows the doped regions of a *silicon-controlled switch* (SCS). Now an external lead is connected to each doped region. Visualize the device separated in two halves (Fig. 15-11b). Therefore, it's equivalent to

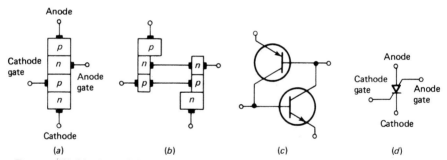

Figure 15-11 SCS. *(a)* Structure. *(b)* Equivalent structure. *(c)* Equivalent circuit. *(d)* Schematic symbol.

a latch with access to both bases (Fig. 15-11c). A forward-bias trigger on either base will close the SCS. Likewise, a reverse-bias trigger on either base will open the device.

Figure 15-11d shows the schematic symbol for an SCS. The lower gate is called the *cathode gate;* the upper gate is the *anode gate.* The SCS is a low-power device compared to the SCR. It handles currents in milliamperes rather than amperes.

324

15-5 BIDIRECTIONAL THYRISTORS

Up to now, all devices have been *unidirectional* because current was in only one direction. This section looks at *bidirectional* thyristors, where current can be in either direction.

Diac

The *diac* can have latch current in either direction. The equivalent circuit of a diac is a pair of four-layer diodes in parallel as shown in Fig. 15-12a, ideally the same as the latches in Fig. 15-12b. The diac is nonconducting until the voltage across it tries to exceed the breakover voltage in either direction.

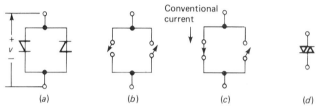

Figure 15-12 Diac. *(a)* Equivalent circuit. *(b)* Latch equivalent. *(c)* Left latch closed. *(d)* Schematic symbol. *(e)* Alternative symbol. *(f)* Another symbol.

For instance, if v has the polarity shown in Fig. 15-12a, the left latch closes when the magnitude of v tries to exceed the breakover voltage. In this case, the left latch closes as shown in Fig. 15-12c. On the other hand, if the polarity of v is opposite that of Fig. 15-12a, the right latch closes when the magnitude of v tries to exceed the breakover voltage.

Once the diac is conducting, the only way to open it is by low-current dropout; this means reducing the current below the rated holding current of the device. Figure 15-12d shows the schematic symbol of a diac.

Triac

The *triac* acts like two SCRs in parallel as shown in Fig. 15-13a, equivalent to the latches of Fig. 15-13b. Because of this, the triac can control current in either direction. The breakover voltage is usually high, and the normal way to turn on a triac is by applying a forward-bias trigger. If v has the polarity shown in Fig. 15-13a, we have to apply a positive trigger; this closes the left latch. When v has a polarity opposite that of Fig. 15-13a, a negative trigger is needed; it will close the right latch.

325

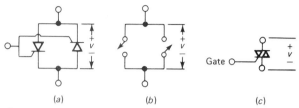

Figure 15-13 Triac. *(a)* Equivalent circuit. *(b)* Latch equivalent. *(c)* Schematic symbol.

Figure 15-13*c* shows the schematic symbol of a triac. The triac is not used as much as the SCR because it cannot handle as much current. Furthermore, there are many more commercially available SCRs than triacs.

15-6 THE UNIJUNCTION TRANSISTOR

The *unijunction transistor* (UJT) has two doped regions with three external leads (Fig. 15-14*a*). It has one emitter and two bases. The emitter is heavily doped, having many holes. The *n* region, however, is lightly doped. For this reason, the resistance between the bases is relatively high, typically 5 to 10 kΩ when the emitter is open. We call this the *interbase resistance,* symbolized R_{BB}.

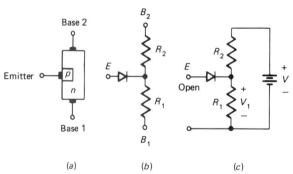

Figure 15-14 Unijunction transistor. *(a)* Structure. *(b)* Equivalent circuit. *(c)* Producing standoff voltage.

Intrinsic standoff ratio

Figure 15-14*b* shows the equivalent circuit of a UJT. The emitter diode drives the junction of two internal resistances, R_1 and R_2. When the emitter diode is nonconducting, R_{BB} is the sum of R_1 and R_2. When a supply voltage

326

is between the two bases as shown in Fig. 15-14c, the voltage across R_1 is given by

$$V_1 = \frac{R_1}{R_1 + R_2} V = \frac{R_1}{R_{BB}} V$$

or
$$V_1 = \eta\, V \tag{15-1}$$

where
$$\eta = \frac{R_1}{R_{BB}}$$

(The Greek letter η is pronounced *eta, e* as the *a* in *face* and *a* as the *a* in *a*bout.)

The quantity η is called the *intrinsic standoff ratio,* which is nothing more than the voltage-divider factor. The typical range of η is from 0.5 to 0.8. For instance, a 2N2646 has an η of 0.65. If this UJT is used in Fig. 15-14c with a supply voltage of 10 V,

$$V_1 = \eta\, V = 0.65 \times 10\,\text{V} = 6.5\,\text{V}$$

How it works

In Fig. 15-14c, V_1 is called the intrinsic standoff voltage because it keeps the emitter diode reverse-biased for all emitter voltages less than V_1. If V_1 equals 6.5 V, we have to apply slightly more than 6.5 V to the emitter to turn on the emitter diode.

What does a UJT do? In Fig. 15-15a, imagine the emitter supply voltage is turned down to zero. Then, the intrinsic standoff voltage reverse-biases the emitter diode. When we increase the emitter supply voltage, v_E increases until it is slightly greater than V_1. This turns on the emitter diode. Since the p region is heavily doped compared to the n region, holes are injected

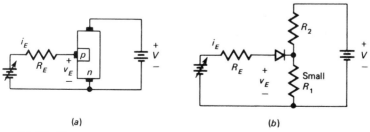

(a)   (b)

Figure 15-15 Turning on a UJT.

in the lower half of the UJT. The light doping of the n region gives these holes a long lifetime. These holes create a p-type inversion layer between the emitter and the lower base (similar to the inversion layer in an E MOSFET).

The flooding of the lower half of the UJT with holes drastically lowers resistance R_1 (Fig. 15-15b). Because R_1 is suddenly much lower in value, v_E suddenly drops to a low value and the emitter current increases.

Latch equivalent circuit

You can remember UJT action by relating it to the latch of Fig. 15-16a. With a positive voltage from B_2 to B_1, a standoff voltage V_1 appears across R_1. This keeps the emitter diode of Q_2 reverse-biased as long as the emitter input voltage is less than the standoff voltage. When the emitter input voltage is slightly greater than the standoff voltage, however, Q_2 turns on and regeneration takes over. This drives both transistors into saturation, ideally shorting the emitter and the lower base.

Figure 15-16b is the schematic symbol for a UJT. The emitter arrow reminds us of the upper emitter in a latch. When the emitter voltage exceeds the standoff voltage, the latch between the emitter and the lower base closes. Ideally, you can visualize a short between E and B_1. To a second approximation, a low voltage called the *emitter saturation voltage,* $V_{E\text{(sat)}}$, appears between E and B_1.

The latch stays closed as long as latch current (emitter current) is greater than the holding current. Data sheets specify a *valley current,* I_V, equivalent to the holding current. For instance, a 2N2646 has an I_V of 6 mA; to hold the latch closed, the emitter current must be greater than 6 mA.

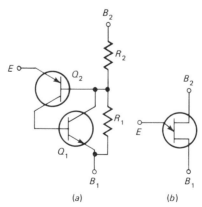

Figure 15-16 UJT. *(a)* Equivalent circuit. *(b)* Schematic symbol.

EXAMPLE 15-6

The 2N4871 of Fig. 15-17 has an η of 0.85. What is the ideal emitter current?

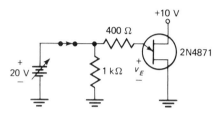

Figure 15-17

SOLUTION

The standoff voltage is

$$V_1 = \eta \, V = 0.85 \times 10 \text{ V} = 8.5 \text{ V}$$

Ideally, v_E must be slightly greater than 8.5 V to turn on the emitter diode and close the latch.

With the input switch closed, 20 V drives the 400-Ω resistor. This is more than enough voltage to overcome the standoff voltage. Therefore, the latch is closed and the emitter current equals

$$I_E \cong \frac{20 \text{ V}}{400 \ \Omega} = 50 \text{ mA}$$

EXAMPLE 15-7

The valley current of a 2N4871 equals 7 mA, and the emitter voltage is 1 V at this point. At what value of emitter-supply voltage does the UJT open in Fig. 15-17?

SOLUTION

As we reduce the emitter-supply voltage, the emitter current decreases. At the point where it equals 7 mA, v_E is 1 V and the latch is about to open. The emitter-supply voltage at this time is

$$V = 1 \text{ V} + (7 \text{ mA} \times 400 \ \Omega) = 3.8 \text{ V}$$

When V is less than 3.8 V, the UJT opens. Then, it will be necessary to raise V above 8.5 V to close the UJT.

329

15-7 UJT CIRCUITS

To get a better idea of how a UJT works and how to use it, here are some practical applications. Figure 15-18a shows a UJT relaxation oscillator. The action is similar to the relaxation oscillator of Example 15-4. The capacitor charges toward V_{CC}, but as soon as its voltage exceeds the standoff voltage, the UJT latches. This discharges the capacitor until low-current dropout occurs. As soon as the UJT opens, the next cycle begins. As a result, we get a sawtooth output.

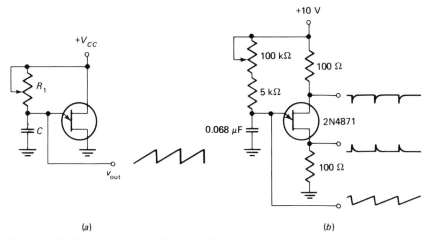

(a)   (b)

Figure 15-18 Relaxation oscillators. *(a)* Sawtooth generator. *(b)* Trigger and sawtooth generator.

If we add a small resistor to each base circuit, we can get three useful outputs: sawtooth waves, positive triggers, and negative triggers as shown in Fig. 15-18b. The triggers appear during the flyback of the sawtooth because the UJT conducts heavily at this time. With the values given in Fig. 15-18b, the frequency is from approximately 50 Hz to 1 kHz, depending on the rheostat value. The width of the triggers (same as flyback time) is in the vicinity of 20 μs.

Sharp trigger pulses out of a UJT relaxation oscillator can be used to trigger an SCR. For instance, Fig. 15-19 shows part of an automobile ignition system. When the distributor points are closed, the UJT and SCR remain open. When the points open, however, the 0.1-μF capacitor charges exponentially. As soon as the capacitor voltage exceeds the standoff voltage, the UJT latch closes and produces the positive trigger as previously described. This trigger is coupled to the gate of the SCR. When the SCR latches

330

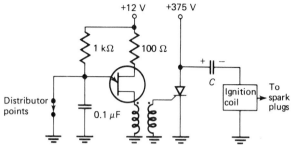

Figure 15-19 Car-ignition system using UJT and SCR.

shut, the positive end of the output capacitor is suddenly grounded. As the output capacitor discharges through the ignition coil, a high-voltage pulse drives one of the spark plugs. When the points again close, the circuit resets itself in preparation for the next cycle.

GLOSSARY

breakover Breakdown followed by regenerative switching that closes the latch.

critical rate of current rise The maximum rate at which current can increase without producing internal hot spots and SCR destruction.

critical rate of voltage rise The maximum rate at which voltage can increase without capacitive currents producing regeneration and premature closing.

forward blocking voltage Equivalent to the breakover voltage of an SCR or triac.

holding current The minimum current that keeps a latch closed.

interbase resistance The resistance between the bases of a UJT when the emitter is open.

latch A switch whose regenerative action can keep it open or closed indefinitely.

relaxation oscillator A circuit that generates a signal whose frequency depends on the time constant of a capacitive or inductive branch.

snubber An *RC* circuit across the SCR to prevent the rate of voltage rise from exceeding the critical value.

331

Transistor Circuit Approximations

transients Temporary changes caused by switches opening or closing. A suddenly changing current, for instance, causes large voltages across inductances. These temporary induced voltages can be fed back to the supply line where they may affect other circuits.

SELF-TESTING REVIEW

Read each of the following and provide the missing words. Answers appear at the beginning of the next question.

1. A latch uses positive feedback, also called _____. It can remain in either of two states, closed or open. One way to close a latch is by _____. Another way is by breakover. Two ways to open a latch are _____ dropout and reverse-bias triggering.

2. *(regeneration, triggering, low-current)* A four-layer diode, also known as a Shockley diode, has _____ doped regions. The only way to close it is by _____. To open it, you have to reduce the current below the _____ current.

3. *(four, breakover, holding)* The silicon-controlled rectifier has a trigger input. SCRs are not intended to break over. The normal way to close an SCR is with a trigger to the _____. The only way to open an SCR is with low-current dropout.

4. *(gate)* To avoid false triggering of the SCR, the anode rate of voltage change must not exceed the _____ rate of voltage rise listed on the data sheet. Switching _____ are the main cause of exceeding this critical rate. Larger SCRs also have a critical rate of _____ rise, which must not be exceeded.

5. *(critical, transients, current)* The diac and triac are _____ thyristors because they can conduct in either direction. The triac acts like two _____ in parallel.

6. *(bidirectional, SCRs)* The unijunction transistor has an intrinsic standoff voltage. To turn it on, the emitter voltage has to be slightly greater than the standoff voltage. To turn it off, the current has to be reduced to slightly less than valley current.

PROBLEMS

15-1. If the 1N5160 of Fig. 15-20a is conducting, for what value of V will it stop conducting (ideally)? If we allow 0.7 V across the diode at the dropout point, what is the value of V where low-current dropout occurs?

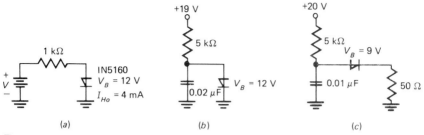

(a) (b) (c)

Figure 15-20

15-2. With a supply of 19 V, it takes the capacitor of Fig. 15-20*b* exactly one time constant to charge to 12 V, the breakover voltage of the diode. If we neglect the voltage across the diode when conducting, what is the frequency of the sawtooth output?

15-3. The current through the 50-Ω resistor of Fig. 15-20*c* is maximum just after the diode latches. Ideally, what is the maximum current? If we allow 1 V across the latched diode, what is the maximum current?

15-4. The 2N4216 of Fig. 15-21*a* has a trigger current of 0.1 mA. Ideally, what value of *V* turns on the SCR? If we allow 0.8 V across the gate-to-ground terminals, what is the value of *V* that turns on the SCR?

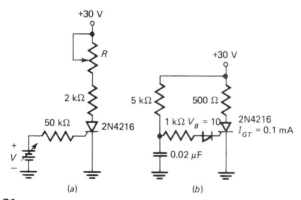

(a) (b)

Figure 15-21

15-5. The four-layer diode of Fig. 15-21*b* has a breakover voltage of 10 V. The SCR has a trigger current of 0.1 mA. Neglecting the voltage across the gate input, what is the current through the four-layer diode just after it breaks over? The current through the 500-Ω resistor after the SCR turns on?

333

Transistor Circuit Approximations

15-6. Figure 15-22a shows an alternative schematic symbol for a diac. The MPT32 diac breaks over when the capacitor voltage reaches 32 V. It takes exactly one time constant for the capacitor to reach this voltage. How long after the switch is closed does the triac turn on? What is the ideal value of gate current when the diac breaks over? The load current after the triac has closed?

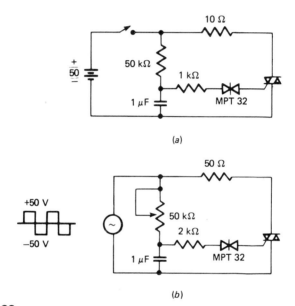

(a)

(b)

Figure 15-22

15-7. The frequency of the square wave in Fig. 15-22b is 10 kHz. It takes exactly one time constant for the capacitor to reach the breakover voltage of the diac. If the MPT32 breaks over at 32 V, what is the ideal value of gate current at the instant the diac breaks over? The ideal load current?

15-8. The UJT of Fig. 15-23a has an η of 0.63. Ideally, what is the value of V that just turns on the UJT? Allowing 0.7 V across the emitter diode, what value of V just turns on the UJT?

15-9. The valley current of the UJT in Fig. 15-23a is 2 mA. If the UJT is latched, we have to reduce V to get low-current dropout. Neglecting the drop across the emitter diode, what value of V just opens the UJT? Allowing 0.7 V across the emitter diode, what is the value of V that just opens the UJT?

334

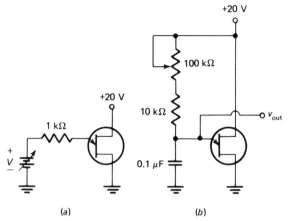

(a)   (b)

Figure 15-23

15-10. The intrinsic standoff ratio of the UJT in Fig. 15-23b is 0.63. Ideally, what are the minimum and maximum output frequencies?

Chapter 16

Pulse and Digital Circuits

The sine wave is the most basic waveform in linear circuits. But when it comes to digital circuits, the rectangular pulse is used, because it produces two-state operation (two points on the load line). This chapter discusses pulse theory and digital circuits.

16-1 THE RECTANGULAR PULSE

A *pulse* is defined as a temporary change in a quantity that is normally constant. In digital electronics the quantity is voltage or current.

Kinds of pulses

A pulse can have any shape. The rectangular pulse of Fig. 16-1*a* increases suddenly to a peak value and later returns suddenly to its *baseline* (where the pulse starts and ends). The sawtooth pulse of Fig. 16-1*b* increases linearly to the peak, then flies back to the baseline. Figure 16-1*c* is a distorted rectangular pulse that starts with an exponential rise and ends with an exponen-

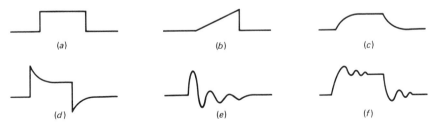

Figure 16-1 Kinds of pulses. *(a)* Rectangular. *(b)* Sawtooth. *(c)* Exponential rise and fall. *(d)* Differentiated edges. *(e)* Ringing. *(f)* Rectangular with ringing.

336

tial decay. Figure 16-1*d* is another distorted rectangular pulse; this time, the pulse begins with a positive step, followed by an exponential decay to a higher level; the pulse ends with a negative step and an exponential return to the baseline. Figure 16-1*e* is a *ringing* pulse, a sine wave whose peaks decay exponentially. In Fig. 16-1*f*, the distorted rectangular pulse has ringing after each transition.

Terminology

Figure 16-2*a* shows some of the terms used to describe the rectangular pulse, the most common of all. The *offset* is the vertical difference between the baseline and zero. The height of the pulse is called the *amplitude;* it equals

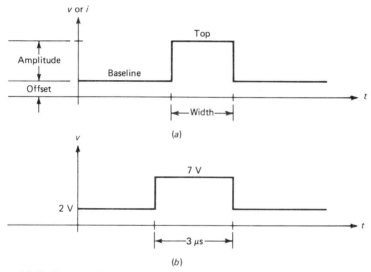

Figure 16-2 Rectangular pulse.

the vertical difference between the top of the pulse and the baseline. The *width* or duration is the time difference between the beginning and end of a pulse.

For instance, if an oscilloscope displays a pulse like Fig. 16-2*b,* we would say the pulse has an offset of 2 V, an amplitude of 5 V, and a width of 3 μs.

Pulse trains

Often, a pulse repeats after a fixed time interval. Figure 16-3*a* shows a pulse that recurs after a period *T.* A string of pulses like this is called a

337

Transistor Circuit Approximations

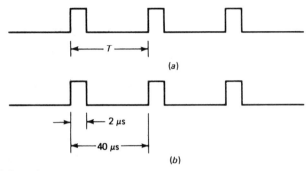

(a)

(b)

Figure 16-3 Pulse train.

pulse train. The frequency of a pulse train, also known as the repetition rate, is the number of pulses per second. Frequency is related to period by the reciprocal formula:

$$f = \frac{1}{T} \qquad \text{(16-1)}$$

For instance, if the period is 40 μs, the frequency or repetition rate is

$$f = \frac{1}{T} = \frac{1}{40 \ \mu s} = 25 \text{ kHz}$$

When the pulse width is much smaller than the period, the pulses appear very narrow. On the other hand, if the width is almost as large as the period, the pulses appear very broad. To pin down how narrow or broad the pulses appear, we use the *duty cycle, D,* given by

$$D = \frac{W}{T} \times 100 \text{ percent} \qquad \text{(16-2)}$$

In a pulse train like Fig. 16-3 *b,* the duty cycle is

$$D = \frac{2 \ \mu s}{40 \ \mu s} \times 100 \text{ percent} = 5 \text{ percent}$$

The duty cycle can vary from 0 to 100 percent. Pulsed laser beams, for example, often have a duty cycle less than 1 percent. On the other hand, the *square wave* is a rectangular pulse train with a duty cycle of 50 percent.

338

Many commercial pulse generators can deliver a rectangular pulse train with a duty cycle from less than 1 percent to more than 90 percent.

16-2 EFFECT OF STRAY CAPACITANCE

Undesirable capacitances exist in any circuit you build. These capacitances have to charge and discharge when the voltage tries to change. This causes distortion of an ideal rectangular pulse, so that the vertical transitions are exponential rather than sudden steps.

Stray capacitance

In hand-wired or printed circuits, each connection between components acts like the plate of a capacitor; the chassis is the other plate. This *stray wiring capacitance* varies from less than 0.2 pF to more than 2 pF per inch of conductor, with the exact value depending on how close the conductor is to ground, its cross-sectional shape, and other factors.

Each resistor in a circuit has *shunt capacitance* distributed over its length. This shunt capacitance depends on the type of resistor (carbon composition, metal-film, etc.), the value of resistance, and other factors. The distributed capacitance of resistors varies from less than 0.003 pF to more than 1 pF.

Bipolar transistors, FETs, and other active devices have internal capacitances. Integrated circuits also have unwanted capacitances, including *parasitic* capacitance between components and the substrate. Therefore, whether discrete or integrated circuits are involved, stray capacitances exist between each signal path and ground.

Lag networks

You can apply Thevenin's theorem to the output of an active device to get an *RC lag network* like Fig. 16-4a. R is the output or Thevenin resistance of the active circuit, and C is the total of all stray and internal capacitances between the output and ground. When the active device has a rectangular output, the capacitance distorts the leading and trailing edges of this pulse. If the capacitance were zero, the final output v_{out} would also be rectangular. But with capacitance in the circuit, the leading and trailing edges become exponential as shown in Fig. 16-4b.

When the source steps positively as in Fig. 16-4a, the capacitance begins to charge to the higher voltage. As the capacitor charges, v_{out} rises exponentially toward V_{top}. This exponential rise is finished after approximately five time constants. Similarly, on the trailing edge of the pulse, the capacitor

339

Transistor Circuit Approximations

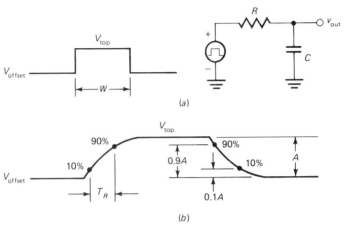

Figure 16-4 *(a)* RC lag network. *(b)* Output.

has to discharge. As it does, v_{out} decreases exponentially to the baseline. The exponential decay is finished after approximately five time constants.

Rise time

In Fig. 16-4 *b*, the 10-percent point is where the voltage has increased to $0.1A$ above the baseline; the 90-percent point is where the voltage is $0.9A$ above the baseline. *Rise time, T_R,* is defined as the time difference between the 90-percent and 10-percent points on the leading edge. Because it's easily measured with an oscilloscope, rise time has become one of the standard specifications used in pulse circuits. For instance, some expensive pulse generators can put out rectangular pulses with rise times of less than 5 ns. Inexpensive pulse generators have rise times in the vicinity of 50 ns.

Fall time, or decay time, is the time it takes for the trailing edge to decrease from the 90-percent point to the 10-percent point. In most applications the *RC* time constant is equal for the leading and trailing edges. For this reason, rise time and fall time are equal.

Rise time and time constant

By setting up the equation for the exponential rise in Fig. 16-4 *b*, we can prove the 10-percent point occurs when $t = 0.1\ \tau$ and the 90-percent point when $t = 2.3\ \tau$, where τ equals the *RC* time constant. Therefore,

$$T_R = 2.3\ \tau - 0.1\ \tau$$

or
$$T_R = 2.2\ \tau \qquad \text{(16-3)}$$

340

For instance, if a lag network has $R = 200$ kΩ and $C = 1$ pF, then $\tau = 200$ ns and

$$T_R = 2.2 \, \tau = 2.2 \times 200 \text{ ns} = 440 \text{ ns}$$

16-3 WAVESHAPING CIRCUITS

Linear waveshaping means using linear components (resistors, inductors, and capacitors) to derive other pulse shapes from an input pulse. This section is about two basic waveshaping circuits.

RC differentiator

Figure 16-5a shows an *RC* circuit. When the time constant is much longer than the period, the pulse train is coupled to the output. But when the time constant is very short, the waveshape is distorted when it reaches the

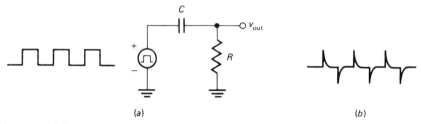

(a)   (b)

Figure 16-5 RC differentiator. *(a)* Circuit. *(b)* Output.

output. Here's why: Suppose the time constant is much shorter than the pulse width. On the leading edge of the pulse, the capacitor voltage cannot change instantaneously; therefore, a voltage step V is coupled to the output. Immediately following this step, however, the capacitor begins to charge exponentially. After approximately five time constants, the capacitor is fully charged and the output voltage has decayed to zero as shown in Fig. 16-5b.

The action on the trailing edge of the input pulse is similar. The capacitor couples the full negative step. Then as it charges in the opposite direction the output voltage decays back to the baseline.

An *RC* circuit like Fig. 16-5a whose time constant is much shorter than the pulse width is called an *RC differentiator.* The smaller the time constant, the narrower are the spikes in Fig. 16-5b. As a guide, keep the time constant less than one-tenth of the pulse width to get narrow output spikes. Differentiating a rectangular pulse is commonly done when you need narrow triggers to drive another circuit.

RC integrator

Figure 16-6*a* shows an *RC integrator,* a lag circuit whose time constant is much longer than the pulse width. Because of the long time constant, the capacitor never fully charges or discharges. In fact, all we see across the output is the earliest part of an exponential charge or discharge. This is why the output looks like the triangular wave of Fig. 16-6*b*. As a guide, the *RC* integrator should have a time constant at least ten times longer than the pulse width.

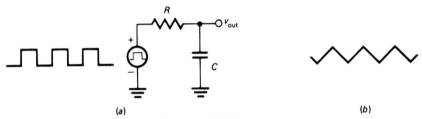

(a)   (b)

Figure 16-6 RC integrator. *(a)* Circuit. *(b)* Output.

In Fig. 16-6*b*, the rising part of the triangular waveform is called a *positive ramp;* the falling part is a *negative ramp.* The longer the time constant is, the more linear these ramps are. But with a longer time constant, we get less amplitude because only the earliest part of an exponential rise or decay is being used.

One way to have a very long time constant and still get large output is with an op-amp integrator like Fig. 16-7*a*. Because of the virtual ground,

$$I_{\text{IN}} = \frac{V_{\text{IN}}}{R}$$

This input current flows to the capacitor and results in this rate of output voltage change:

$$\frac{\Delta V_{\text{OUT}}}{\Delta t} = \frac{-I_{\text{IN}}}{C}$$

or

$$\frac{\Delta V_{\text{OUT}}}{\Delta t} = \frac{-V_{\text{IN}}}{RC}$$

where the minus sign indicates phase inversion.

The main use of an op-amp integrator like Fig. 16-7*a* is to generate a

342

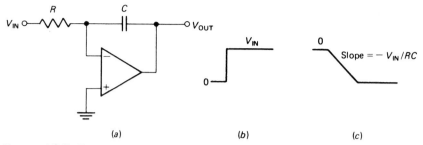

Figure 16-7 Op-amp integrator.

ramp. To get a ramp output, we need to use a step-voltage input like Fig. 16-7b. This produces a constant input current, which forces the output to slew negatively as shown in Fig. 16-7c. Since the slope equals $-V_{IN}/RC$, we can control the rate of change by varying V_{IN} or by using different RC values. Linear ramps like this are used in digital voltmeters, oscilloscopes, and many other applications.

Compensated voltage divider

A voltage divider is often used at the front end of measuring instruments to prevent large signals from overdriving the system. Figure 16-8a shows a voltage divider, including the stray capacitance C_2 across R_2. Because of this stray capacitance, a rectangular input pulse is distorted by the rise and fall times as shown. To overcome the effects of the lag network, we can add a compensating capacitor C_1 as shown in Fig. 16-8b. This shunts ac current around R_1 and compensates for the shunting effect of C_2 around R_2.

Perfect compensation occurs when the ratio of resistances equals the ratio of capacitive reactances:

$$\frac{R_2}{R_1} = \frac{X_{C2}}{X_{C1}} = \frac{\frac{1}{2}\pi fC_2}{\frac{1}{2}\pi fC_1}$$

or
$$C_1 = \frac{R_2}{R_1} C_2 \tag{16-4}$$

C_1 is usually a trimmer capacitor to allow a precise compensation of C_2, which is a stray capacitance. Commonly, C_1 is adjusted using a square-wave input. When C_1 is too large, we get a differentiated output (Fig. 16-8c). When it's too small, we get an integrated output. And when it's perfect, we get a square-wave output as shown.

343

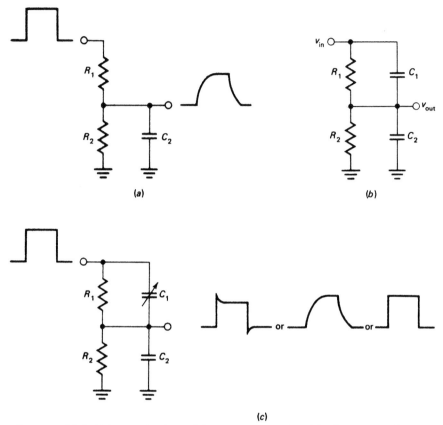

Figure 16-8 Voltage divider. *(a)* Uncompensated. *(b)* Compensated. *(c)* Degrees of compensation.

16-4 LOGIC CIRCUITS

Almost all digital computers and systems are based on *binary* (two-state) operation. Because of this, all input and output voltages are either low or high. This type of operation is ideally suited to digital integrated circuits, which are now in widespread use.

Saturated logic

Some of the circuits in digital electronics are called *logic* circuits because they duplicate processes of the mind. The simplest way to design logic circuits is the cutoff-saturation approach described earlier; the idea is to operate all transistors at cutoff or saturation. When logic circuits use saturation as one point on the load line, they are called *saturated* logic circuits, or simply *saturated logic.*

Although it's the simplest, saturated logic has a speed limitation caused by saturating the transistors. To understand the problem, look at Fig. 16-9a, where a rectangular pulse drives the base of a switching transistor. Before

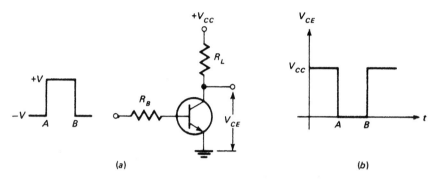

(a)   (b)

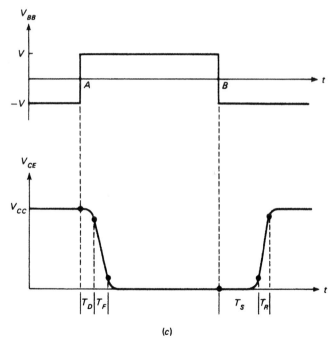

(c)

Figure 16-9 Saturated logic. *(a)* Example. *(b)* Ideal output. *(c)* Input and output.

Transistor Circuit Approximations

point A in time, the transistor is cut off and the collector voltage is V_{CC}. Suddenly, at point A the base voltage changes from $-V$ to $+V$. If we ignore all capacitances, the transistor will immediately go into saturation, causing the collector voltage to drop to zero as shown in Fig. 16-9b. The transistor remains saturated until point B in time, when the base voltage returns to its baseline and the collector voltage jumps back to $+V_{CC}$.

The vertical transitions of Fig. 16-9b are only approximations, because we ignored stray capacitances and other factors. Here is what actually happens when we take everything into account: Figure 16-9c shows the base and collector pulses. The transitions have been exaggerated to bring out certain ideas. As you see, even though the base voltage switches suddenly, the collector voltage cannot immediately follow because of stray capacitances and charge-storage effects.

Specifically, the leading edge of the collector pulse is slowed down by these two effects:

1. *Delay time*, T_D, is the amount of time that elapses between point A and the 90-percent point. Two factors contribute to this delay: the time needed for carriers to pass from the emitter to the base, and the base-emitter capacitance.
2. *Fall time*, T_F, is the time difference between the 90-percent and 10-percent points. Stray capacitance (including internal transistor capacitance) between the collector and ground forms a lag network responsible for the fall time.

The trailing edge of the collector pulse is slowed down by the following effects:

1. Storage time, T_S, (also called the saturation delay time) is the time between point B and the 10-percent point. Storage time is the result of charge storage in the base region of a saturated transistor. The more heavily saturated the transistor, the longer the storage time. Until all stored carriers have been removed from the base region, the transistor remains saturated even though the base voltage has switched off.
2. *Rise time*, T_R, is the time between the 10-percent and 90-percent points. Like fall time, rise time is the result of a lag network formed by stray and internal transistor capacitances that appear between the collector and ground.

The foregoing times set an upper limit on how fast we can switch a transistor on and off. For instance, suppose we have the following times: $T_D = 5$ ns, $T_F = 10$ ns, $T_S = 40$ ns, and $T_R = 10$ ns. The sum of these is 65 ns.

346

Therefore, to turn a transistor on and off, we need at least 65 ns. Especially important, storage time T_S is always the longest. If we could eliminate it, the total switching time would drop to 25 ns, an improvement of more than 100 percent.

Despite the storage-time problem, saturated logic is the most widely used logic because there are many applications where speed is not necessary.

Nonsaturated logic

One way to speed up a logic circuit is by preventing transistor saturation. Circuits in this category are called *nonsaturated logic.* By avoiding transistor saturation, we can prevent excessive charge storage and almost eliminate the storage time T_S that slows down saturated logic circuits. Examples of nonsaturated logic will be given later.

16-5 DIGITAL INTEGRATED CIRCUITS

The earlier part of this book has already discussed some digital ICs. This section presents new examples, as well as summarizing what has gone before.

TTL

Transistor-transistor logic (TTL) became commercially available in 1964. Since then, it has become the most popular bipolar type of IC. Figure 16-10 is an example of a TTL circuit. Notice the *multiple-emitter* input transistor

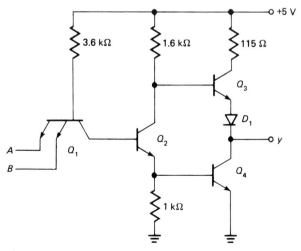

Figure 16-10 Example of a TTL circuit.

and the *totem-pole* output transistors. The transistors operate between cutoff and saturation; therefore, TTL is a saturated type of logic.

Here's the action: When A and B are both low (ideally 0 V to ground), Q_1 saturates; this cuts off Q_2 and Q_4. Then, Q_3 acts like an emitter follower and couples a high voltage to the output. This is the first entry of Table 16-1. When A is low and B is high, Q_1 remains saturated and the final output is high. Likewise, when A is high and B is low, the final output is high. Finally, when A and B are both high, forward current through the Q_1 collector diode forces Q_2 and Q_4 to saturate. Because Q_4 is saturated, the output voltage is low, as shown in Table 16-4.

TABLE 16-1 NAND GATE

A	B	y
Low	Low	High
Low	High	High
High	Low	High
High	High	Low

The circuit of Fig. 16-10 is called a NAND *gate*. By varying the design, manufacturers can produce OR gates, AND gates, and a variety of other basic logic circuits. TTL circuits are widely used; more than 75 percent of all bipolar ICs are in the TTL family.

Schottky TTL

Because TTL is a saturated logic, the switching transistors have a storage-time problem that slows down the action. One way to overcome this is with *Schottky TTL*, a variation of standard TTL where a Schottky diode is connected between the collector and the base as shown in Fig. 16-11. The typical forward voltage of a Schottky diode is only 0.4 V. When a transistor

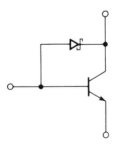

Figure 16-11 Schottky TTL.

348

saturates, its collector diode becomes forward-biased. By including the Schottky diode, we prevent transistor saturation because the Schottky diode turns on before the transistor can fully saturate. Since the Schottky diode has no charge storage, the storage time of the combination approaches zero.

By optimizing the design for minimum current drain, manufacturers are producing *low-power* Schottky TTL, a very popular variation of TTL because of its high speed and low power consumption.

ECL

Emitter-coupled logic (ECL) is the fastest bipolar type of logic. An ECL circuit uses diff-amp stages whose transistors stay well out of saturation. Figure 16-12 is an example of an ECL circuit. When *A* and *B* are both

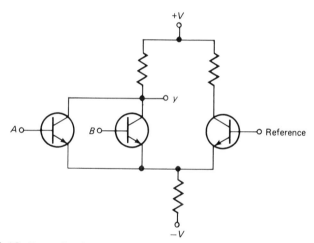

Figure 16-12 Example of ECL circuit.

low, the two transistors on the left are off and the output *y* is high. If either *A* or *B* is high, the output is low. And when *A* and *B* are both high, *y* is low. Table 16-2 summarizes the input/output possibilities. A circuit like this is called a *NOR gate*.

TABLE 16-2   NOR GATE

A	B	y
Low	Low	High
Low	High	Low
High	Low	Low
High	High	Low

349

By varying the design, manufacturers can produce OR gates, AND gates, and other basic ECL circuits. The main advantage of this nonsaturated logic is its high speed.

CMOS

We discussed this earlier. As you recall, it uses p-channel and n-channel enhancement MOSFETs. The main advantage is its extremely low power consumption. This is why CMOS is preferred in very low-power applications like electronic wrist watches, pocket calculators, space vehicles, etc.

PMOS and NMOS

As mentioned earlier, p-channel MOS is a logic using p-channel drivers and loads. More popular nowadays is n-channel MOS because its n-channel drivers and loads can switch on and off faster than PMOS. Currently, NMOS is used in large-scale integration such as memories, microprocessors, etc.

Summary of basic logic circuits

The digits 0 and 1 are normally used to represent low and high voltages. For example, instead of saying A is low and B is high, we can say $A = 0$ and $B = 1$. Figure 16-13 summarizes the basic logic circuits of digital electronics. All of these were discussed earlier. Figure 16-13a is an OR gate. As shown in the table, it has a high output when A or B or both are high. The AND gate of Fig. 16-13b has a high output only when A and B are both high. The NOR gate of Fig. 16-13c has a high output only when A and B are both low. The NAND gate (Fig. 16-13d) has a low output only when A and B are both high.

By combining these basic logic circuits, we can build the complex circuits

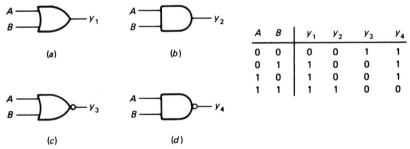

A	B	Y_1	Y_2	Y_3	Y_4
0	0	0	0	1	1
0	1	1	0	0	1
1	0	1	0	0	1
1	1	1	1	0	0

Figure 16-13 Logic gates. *(a)* OR. *(b)* AND. *(c)* NOR. *(d)* NAND.

350

found in digital computers. In other words, with combinations of the basic logic circuits shown in Fig. 16-13, computers can do arithmetic and process data in all kinds of ways. Books specializing in digital electronics and computers can show you how it's done.

16-6 MULTIVIBRATORS

A *multivibrator* is a regenerative circuit with two active devices, designed so that one device conducts while the other cuts off. Multivibrators can store numbers, count pulses, generate rectangular waves, and perform other functions in a digital computer.

Astable multivibrator

The *astable* multivibrator has two states but is stable in neither. In other words, it will remain in one state temporarily, and then switch to the other state. Here it remains for a while before switching back to the original state. This continuous switching back and forth produces a rectangular wave with fast rise time and fall time. Because no input signal is required to get an output, the astable multivibrator is sometimes called a *free-running* multivibrator.

One way to build an astable multivibrator is with a *555 timer,* an IC with dozens of different functions. Figure 16-14 shows a 555 timer connected for astable operation. It free-runs at a frequency determined by R_A, R_B, and C. The duration of the high output state is

$$T_H = 0.693(R_A + R_B)C$$

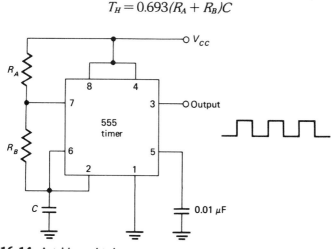

Figure 16-14 Astable multivibrator.

351

and the duration of the low output state is

$$T_L = 0.693R_BC$$

Therefore, the period of the rectangular output is

$$T = 0.693(R_A + 2R_B)C$$

and the frequency is

$$f = \frac{1.44}{(R_A + 2R_B)C} \qquad \text{(16-5)}$$

The duty cycle is given by

$$D = \frac{T_H}{T} \times 100 \text{ percent}$$

or

$$D = \frac{R_A + R_B}{R_A + 2R_B} \times 100 \text{ percent} \qquad \text{(16-6)}$$

Monostable multivibrator

The *monostable* multivibrator is stable in one state but unstable in the other. When triggered, it goes from the stable state to the unstable state. It remains in the unstable state temporarily and then returns to the stable state. Figure 16-15 illustrates the idea of a monostable multivibrator. A sharp positive trigger at point A in time causes the output to go high; the output stays high temporarily and then returns to the baseline. The pulse width is usually controlled by a capacitor. The output will stay low indefinitely until another input trigger comes in (point B). Again, we get an output pulse of width W. Each time a trigger hits the monostable multivibrator, we get one output

Figure 16-15 Waveforms for monostable multivibrator.

pulse. This is why a monostable multivibrator is often a called a *one-shot* multivibrator (or simply a one-shot).

The 555 timer can be connected for monostable operation as shown in Fig. 16-16. Because of its internal design, the 555 requires *negative* input

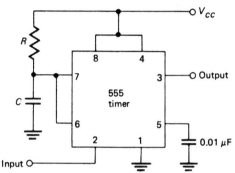

Figure 16-16 Monostable multivibrator.

triggers. Each time a negative trigger hits pin 2, a single pulse comes out of pin 3. The width of this pulse is determined by R and C, and is given by

$$W = 1.1RC \qquad\qquad (16\text{-}7)$$

Bistable multivibrator

The *bistable multivibrator* has two stable states. It can remain in either state indefinitely. When triggered, it goes from one state to the other, where it remains until receiving the next trigger. Figure 16-17 shows the input triggers and the output wave. As you see, each time a trigger arrives, the output

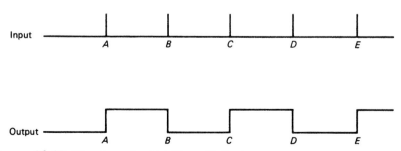

Figure 16-17 Waveforms for bistable multivibrator.

353

changes to the other state. Notice that it takes two triggers to complete one output cycle. This divide-by-two property has made the bistable multivibrator very important in electronic counters.

Figure 16-18 shows one way to build a bistable multivibrator (also called a *flip-flop*) using a 7470, a digital IC. The 7470 is designed to trigger on the rising edge of each input pulse. This means it triggers at points *A*, *C*, and so on. As it does, the output changes to the opposite state. Notice that it takes two input pulses to get one output pulse.

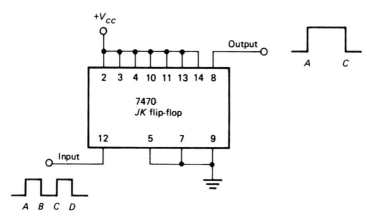

Figure 16-18 Bistable multivibrator.

GLOSSARY

astable multivibrator Also known as a free-running multivibrator, this regenerative circuit is a type of relaxation oscillator whose rectangular output has a period controlled by a timing capacitor.

bistable multivibrator Also called a flip-flop, this circuit has two stable states. Each time it receives a trigger, it goes to the opposite state. Its divide-by-two property is especially important in electronic counters.

duty cycle The percentage ratio of pulse width to period.

monostable multivibrator Also known as a one-shot, this circuit delivers one output pulse for each input trigger.

nonsaturated logic Digital circuits whose transistors do not use saturation as one of the points on the load line.

ramp A linearly changing voltage.

rise time The time it takes for a pulse to rise from the 10-percent point to the 90-percent point.

354

saturated logic Digital circuits whose transistors use saturation as one of the points on the load line.

square wave A rectangular pulse train with a duty cycle of 50 percent.

storage time Also called saturation delay time, storage time is the time it takes to remove all stored charges from the base region of a saturated transistor whose base drive is suddenly turned off.

SELF-TESTING REVIEW

Read each of the following and provide the missing words. Answers appear at the beginning of the next question.

1. Rise time is the time it takes for a pulse to rise from its _____ point to its _____ point. Fall or decay time usually equals rise time because the circuit time constant remains the same in most applications.

2. *(10-percent, 90-percent)* When driven by a rectangular input, an *RC* differentiator produces narrow spikes. To get these, the time constant must be much _____ than the pulse width. The *RC* integrator driven by a rectangular input produces positive and negative _____.

3. *(smaller, ramps)* The simplest way to design logic circuits is the cutoff-saturation approach. Circuits like this are called _____ logic. The main disadvantage of this method is the storage time or saturation delay time, which limits the switching speed.

4. *(saturated)* One way to speed up a logic circuit is by preventing transistor saturation. Circuits in this category are called _____ logic.

5. *(nonsaturated)* TTL is widely used. Standard TTL is an example of saturated logic. One way to get nonsaturated logic is by connecting a _____ diode between the collector and base; when optimized for low-power consumption, this nonsaturated logic is called low-power _____ TTL.

6. *(Schottky, Schottky)* CMOS uses *p*-channel and _____ devices. The main advantage of CMOS is its very low _____ consumption. It's used in electronic wrist watches, pocket calculators, and so on.

7. *(n-channel, power)* NMOS uses *n*-channel drivers and loads. It is used for _____ integration of memories, microprocessors, etc.

8. *(large-scale)* The astable multivibrator generates a rectangular pulse train.

355

The monostable multivibrator produces one rectangular pulse for each trigger. The bistable multivibrator has the divide-by-two property.

PROBLEMS

16-1. A pulse train has a pulse width of 5 ns and a period of 64 ns. What is the duty cycle?

16-2. A rectangular wave is high for 30 μs and low for 20 ns. What is the duty cycle?

16-3. An RC lag network has $\tau = 30$ μs. What is the rise time?

16-4. In Fig. 16-7a, $V_{IN} = 8$ V, $R = 50$ kΩ, and $C = 1000$ pF. What does the output slope equal?

16-5. The voltage divider of Fig. 16-8b has $R_1 = 90$ kΩ, $R_2 = 10$ kΩ, and $C_2 = 50$ pF. What value should the compensating capacitor C_1 have?

16-6. The astable multivibrator of Fig. 16-14 has these values: $R_A = 1$ kΩ, $R_B = 10$ kΩ, and $C = 0.1$ μF. Calculate the frequency and the duty cycle.

16-7. The monostable multivibrator shown in Fig. 16-16 has these values: $R = 1$ kΩ and $C = 0.01$ μF. What is the width of the output pulse?

16-8. The input frequency to a bistable multivibrator is 330 Hz. What is the output frequency?

Answers to Odd-Numbered Problems

CHAPTER 2

2–1. 0.3 A **2–3.** *(a)* 0.293 A; *(b)* 0.318 A **2–5.** 0.05 A **2–7.** 0; −14 V **2–9.** 0.93 mA **2–11.** 35.4 μA

CHAPTER 3

3–1. 10 mA **3–3.** *(a)* 2.9 mA; *(b)* 14.4 V **3–5.** Ideally, a negative half-wave signal with a peak of −10 V; to a second approximation, the negative peak is still −10 V, but the positive peak is clipped at 0.4 V. **3–7.** 2.52 MHz; 5.03 MHz **3–9.** 50.7 V; 51.7 V **3–11.** 6 mA; 14 mA **3–13.** *(a)* 8.33 kΩ; *(b)* 3.57 kΩ

CHAPTER 4

4–1. 148 V **4–3.** 23.6 V; 74 V **4–5.** 32.2 V **4–7.** 17 A **4–9.** Approximately 29 V **4–11.** *(a)* 29.7 V; *(b)* 13.8 mA; *(c)* 11.3 mA; *(d)* 110 mA **4–13.** 1527 V **4–15.** 200 Ω; 0; 98 mA

CHAPTER 5

5–1. 0.49 mA **5–3.** 0.49 mA dc and 0.06 mA rms ac **5–5.** *(a)* 500 Ω; *(b)* 90.9 Ω; *(c)* 20.8 Ω; *(d)* 3.76 Ω **5–7.** 0.524 mA; 47.7 Ω **5–9.** 0.0371 mA **5–11.** 0.524 mA; 0.524 mA dc and 0.21 mA rms ac

357

CHAPTER 6

6–1. 0.98; 50   **6–3.** 0.99; 0.997   **6–5.** 0.495 mA; −10.1 V   **6–7.** 0.495 mA; 50.5 Ω   **6–9.** 3.67 V; 0.152 V rms; 3.67 V dc; and 0.152 V rms ac

CHAPTER 7

7–1. 1.93 mA; 11.9 V   **7–3.** 4.9 mA; 1.239 MΩ   **7–5.** 16.1 mA; 9.71 V   **7–7.** 2.89 mA; 7.1 V; 2.17 V   **7–9.** 5.79 mA; 13.4 V   **7–11.** 1.68 mA; 12.1 V; 5.54 V   **7–13.** 4.19 mA; −13.7 V   **7–15.** 17.8 V; 10.7 V **7–17.** 4.37 kΩ; 240 mV rms   **7–19.** 0.194 V rms

CHAPTER 8

8–1. 0.966   **8–3.** 0.993; 1.45 kΩ; 1.01 kΩ   **8–5.** 121; 198   **8–7.** 163; 1.03 MΩ

CHAPTER 9

9–1. 1.95 A; 29.9 V; 13.7 A; 34.9 V   **9–3.** 36.8 V; no   **9–5.** Yes   **9–7.** 39.2 W; approximately 4.89 W   **9–9.** 132 W; 66 W   **9–11.** 40.7 W; 20.3 W   **9–13.** 71.4; 3.92 Ω   **9–15.** 100°C   **9–17.** 99.3 mW   **9–19.** 125°C

CHAPTER 10

10–1. *(a)* AB; *(b)* B; *(c)* A; *(d)* C   **10–3.** 0.65 mA   **10–5.** 36.7 mA   **10–7.** 3.125 W; 0.625 W   **10–9.** 10 Ω; 0.988; 49.4   **10–11.** 39.4 A

CHAPTER 11

11–1. 42.7   **11–3.** 10 kΩ; 250; 25   **11–5.** Approximately 1.5 mA   **11–7.** 7.5 mA; 300   **11–9.** 400; 400 mV   **11–11.** 300 MHz; 1.43 MHz   **11–13.** 40; 40; 1 kΩ; 1 MΩ

CHAPTER 12

12–1. 1.5(10^{11}) Ω   **12–3.** 4.32 mA   **12–5.** 2500 μS   **12–7.** 250 Ω **12–9.** 18.6 V; 2.22 V   **12–11.** 3 mA; 15 V   **12–13.** 0.849; 2 MΩ   **12–15.** 75 mV   **12–17.** 53 mV; 8 V

CHAPTER 13

13–1. $2(10^{11})$ Ω   **13–3.** 14.2 V; 8 V   **13–5.** *(a)* 10 V; *(b)* 10 V; *(c)* 1 V; *(d)* the gate voltage is less than the threshold voltage.   **13–7.** 1.5 V   **13–9.** 20 V; 4 mA

CHAPTER 14

14–1. 20 μV   **14–3.** 0.995 MHz   **14–5.** 221   **14–7.** 2.86 kHz   **14–9.** 2 kΩ; 34

CHAPTER 15

15–1. 4 V; 4.7 V   **15–3.** 0.18 A; 0.16 A   **15–5.** 10 mA; 60 mA   **15–7.** 16 mA; 1 A   **15–9.** 2 V; 2.7 V

CHAPTER 16

16–1. 0.0781, or 7.81 percent   **16–3.** 66 μs   **16–5.** 5.56 pF   **16–7.** 11 μs

Appendix

Abbreviations and Symbols

A	voltage gain	Hz	hertz (cps)
AM	amplitude modulation	i	total current
		i_b	ac base current
B	bandwidth	i_B	total base current
C_c	collector-diode capacitance	I_B	dc base current
		i_c	ac collector current
C_e	emitter-diode capacitance	i_C	total collector current
C_E	emitter bypass capacitance	I_C	dc collector current
CdS	cadmium sulfide	I_{CBO}	collector-base leakage current with open emitter
DE MOSFET	depletion-enhancement MOSFET	I_{CEO}	collector-emitter leakage current with open base
E MOSFET	enhancement MOSFET	I_{CO}	same as I_{CBO}
$f\alpha$	alpha cutoff frequency	$I_{C(\text{sat})}$	collector saturation current
f_β	beta cutoff frequency	i_d	ac drain current
f_T	gain-bandwidth product	I_D	dc drain current
		I_{DSS}	shorted-gate drain current
FET	field-effect transistor	i_e	ac emitter
g_m	transconductance	i_E	total emitter current
GCS	gate-controlled switch	I_E	dc emitter current

361

Transistor Circuit Approximations

I_V	valley current	R_S	external source resistance of JFET
IC	integrated circuit		
IGFET	insulated-gate FET	r_Z	zener resistance
JFET	junction FET	SCR	silicon-controlled rectifier
K	beta sensitivity		
laser	*l*ight *a*mplification by *s*imulated *e*mission of *r*adiation	SCS	silicon-controlled switch
		UJT	unijunction transistor
LED	light-emitting diode	v_C	total collector-ground voltage
LSI	large-scale integration		
		v_{CB}	total collector-base voltage
MOSFET	metal-oxide semiconductor FET	v_{CE}	total collector-emitter voltage
η	intrinsic standoff ratio of UJT	V_{BE}	dc base-emitter voltage
p-i-n	*p* type-intrinsic-*n* type	V_C	dc collector-ground voltage
r'_b	base spreading resistance	V_{CC}	collector supply voltage
r_B	bulk resistance	V_{CE}	dc collector-emitter voltage
R_B	external base resistance		
R_{BB}	interbase base resistance of UJT	V_{DD}	drain supply voltage
		V_E	dc emitter-ground voltage
r'_e	emitter ac resistance	V_{EE}	emitter supply voltage
r_E	unbypassed emitter resistance	V_{GG}	gate supply voltage
R_E	external emitter resistance	$V_{GS(off)}$	gate cutoff voltage
r_L	ac load resistance	$V_{GS(th)}$	gate threshold voltage
R_L	dc load resistance	V_P	pinchoff voltage
r_S	unbypassed source resistance of JFET	V_Z	zener voltage

Greek Symbols and Letters

α	ac alpha	β_0	low-frequency beta
α_{dc}	dc alpha	$\Delta\beta$	change in beta
β	ac beta	Δi	change in current
β_{dc}	dc beta	Δv	change in voltage
β_{DP}	Darlington-pair beta	μ	amplification factor

362

References for Further Reading

Boylestad, R., and L. Nashelsky: *Electronic Devices and Circuit Theory,* 2d ed., Prentice-Hall, Inc., Englewood Cliffs, N.J., 1978.

Deboo, G. J., and C. N. Burrous: *Integrated Circuits and Semiconductor Devices,* 2d ed., McGraw-Hill Book Company, New York, 1977.

Graeme, J. G., G. E. Tobey, and L. P. Huelsman: *Operational Amplifiers,* McGraw-Hill Book Company, New York, 1971.

Kiver, M. S.: *Transistor and Integrated Electronics,* 4th ed., McGraw-Hill Book Company, New York, 1972.

Lurch, E. N.: *Fundamentals of Electronics,* 2d ed., John Wiley & Sons, Inc., New York, 1971.

Malvino, A. P.: *Electronic Principles,* 2d ed., McGraw-Hill Book Company, New York, 1979.

Malvino, A. P., and D. P. Leach: *Digital Principles and Applications,* 2d ed., McGraw-Hill Book Company, New York, 1975.

Motorola: *The Semiconductor Data Book,* 5th ed., Motorola Semiconductor Products, Inc., Phoenix, Az., 1970.

RCA: *Thyristors, Rectifiers, and Other Diodes,* RCA Solid-State Division, Somerville, N.J., 1972.

Smith, R. J.: *Circuits, Devices, and Systems,* 3d ed., John Wiley & Sons, Inc., New York, 1976.

Stern, L.: *Fundamentals of Integrated Circuits,* Hayden Book Company, Inc., New York, 1968.

Tocci, R. J.: *Fundamentals of Electronic Devices,* 2d ed., Charles E. Merrill Publishing Company, Columbus, Ohio, 1975.

363

Index

Index

367

Index

369

Index

370